DYNAMIC PROJECT MANAGEMENT

A Practical Guide for Managers and Engineers

DEBORAH S. KEZSBOM
President
MRA Management Resources, Inc.
Englewood Cliffs, New Jersey

DONALD L. SCHILLING
Herbert Kayser Professor of
Electrical Engineering and Computer Science
The City College of New York
New York, New York

KATHERINE A. EDWARD
Member of the Technical Staff
AT&T Bell Laboratories
Middletown, New Jersey

WILEY

A WILEY-INTERSCIENCE PUBLICATION

JOHN WILEY & SONS

New York · Chichester · Brisbane · Toronto · Singapore

To those who loved and supported us through this "project effort"

Copyright © 1989 by John Wiley & Sons, Inc.

All rights reserved. Published simultaneously in Canada.

Reproduction or translation of any part of this work
beyond that permitted by Section 107 or 108 of the
1976 United States Copyright Act without the permission
of the copyright owner is unlawful. Requests for
permission or further information should be addressed to
the Permissions Department, John Wiley & Sons, Inc.

Library of Congress Cataloging in Publication Data:
Kezsbom, Deborah S.
   Dynamic project management : a practical guide for managers and
engineers / Deborah S. Kezsbom, Donald L. Schilling, Katherine A.
Edward.
     p.   cm.
   "A Wiley-Interscience publication."
   Bibliography: p.
   Includes index.
   ISBN 0-471-85248-1
   1. Industrial project management. I. Schilling, Donald L.
II. Edward, Katherine A. III. Title.
T56.8.K45 1988
658.4'04--dc19
                               88-21979
                                   CIP

Printed in the United States of America
10 9 8 7 6 5 4

PREFACE

This book is about high-tech engineering projects and the people who implement them. It is a comprehensive and practical guide for managing today's competitive and demanding high-tech project environment; of meeting profit goals and motivating the professional of the new decade.

We, ourselves, are a small project team and have brought together our collective experiences within the private and public sectors to share with you, the reader, our understanding of engineering projects and engineering organizations and how to successfully work within them. We draw our knowledge, information, and excitement from the organizations from which we grew and the clients we now serve. What you will read in this book is not merely a compilation of academic research and training. It is an account of what we have practiced and have seen practiced well. To maintain the confidential and proprietary nature of our client's organizations, we have chosen to use what appear to be generic examples, and have taken author's license with our examples and case studies. Direct reference to people or organizational practices are therefore cited as appropriate.

We believe that we have addressed all aspects of project management: the basic and uniquely differentiating characteristics of projects, the differences and difficulties of working within project organizations, the role of the project manager, the skills needed to survive in a multiproject environment, building the project team, dealing with inevitable conflict, responding to the Request for Proposal, and evaluating and selecting computer-based project management information systems.

In Chapter 1, we examine the challenges of high technology today that encouraged managers and project personnel to contemplate the development of more formal project management structures and procedures. We further explain, in Chapter 2, how properly designed organizations can be major contributors to project success and describe the advantages and disadvantages of three widely accepted organizational structures. We deal with the question of project interface and authority, for unless lines of responsibility and reporting relationships are clearly delineated, conflict will result. The importance of proper project planning is covered in Chapter 3. Techniques for planning and determining requirements, schedules, and budgets for the different life-cycle phases of a project are explained. Chapter 4 describes the principal scheduling techniques that we have found to be effective in planning, tracking, and assuring quality. In Chapter 5, guidelines for establishing project control systems to measure fiscal and technical performance are introduced.

Project leadership and the role of the project manager are discussed in Chapter 6. Chapter 7 illustrates how, if assessed and managed properly, conflict can actually be constructive and improve product development. Skills needed to foster good communication within the project environment are presented in Chapter 8. It is essential for managers of today's high-tech projects to constantly work towards establishing a work environment conducive to team building. How to build the project team and identify and overcome barriers to team performance is the subject of Chapter 9.

In Chapter 10, we explain how to respond to a Request for Proposal (RFP) from a government agency, another company, or another division within your own company. And, finally, in Chapter 11, the features and capabilities of several computer-based project management systems are presented, with suggestions for assessing and selecting the right program to match project needs.

Our intent is to create a practical text of work that can be immediately applied by managers and project specialists alike. We also want to promote some philosophies of project management that we believe are not only basic but crucial to project success: philosophies such as integrative project planning, trust, and frequent, supportive organizational communication. For this purpose, we have followed the format of our seminars and have included a variety of exercises, case studies, and examples. The book, in fact, is a product of the hundreds of seminars we have delivered to engineers and technical specialists across the country and represents the knowledge imparted to and gleaned from such dynamic organizations as AT&T, Motorola, 3M, General Instrument, NASA, and other agencies within the Department of Defense.

We encourage our academic colleagues to consider the text as an excellent source of information for a one-semester course in project management. It can be used successfully at both the graduate and undergraduate levels.

The authors would like to acknowledge the help provided by their thousands of students, who, as managers and managers to be, provided us with a sea of examples on how to manage (and how not to manage) projects. In addition we wish to acknowledge to assistance provided by Ms. Sarita Whitehead, Ms. Victoria Benzinger, and Mr. Joseph Saez, who typed, retyped, and typed once again the numerous drafts of this manuscript.

DEBORAH S. KEZSBOM
DONALD L. SCHILLING
KATHERINE A. EDWARD

CONTENTS

CHAPTER 1

THE DYNAMICS OF PROJECT MANAGEMENT

1.1 INTRODUCTION

On July 20, 1969, Americans watched in awe and with pride as the space craft *Eagle* landed in the Sea of Tranquility and humankind took its first step on the moon. To successfully land and return such a vessel to a target destination over 250,000 miles away represented not only a monumental and auspicious scientific accomplishment but also a fantastic *managerial* feat. In response to the scientific and managerial complexities inherent in the Apollo space program, new and unique methods were derived to restructure management and adapt special procedures for better command, control, and communication (C^3). Thus, more than 20 years ago within the Department of Defense, management "celebrated" the birth of a formal approach to manage projects, which served as the forerunner to our modern conceptions of project management. These concepts and procedures continue to be applied and adapted today to more competitive and commercial project efforts.

The development and installation of project management systems within the commercial sectors is not without substantial resistance. Traditional corporate structures were designed to be stable and survive over time. For these reasons, these structures prove time and time again to be incapable of responding to complex, dynamic, and changing project requirements. Instituting formal as well as informal project management systems and procedures into existing traditional organizations represents a major change. And change takes time.

Adopting a project management approach requires a long, hard examination of existing management structures, current planning, scheduling and controling techniques. Most of all, efforts need be focused toward the personnel who are the cornerstone of project efforts.

1.2 WHY "PROJECT" MANAGEMENT?

The challenge of high technology in this decade and the one to come has forced managers and project personnel alike to reconsider and more closely examine their present management techniques. Despite resistance to change, managers have had to contemplate the use of more systematic project management approaches to oversee product development and production.

Not all high-technology organizations may deem it necessary to institute formal project structures and procedures. However, companies that are involved in the development of complex technological products and services find that:

- The *time span* between the initiation of a project and its actual completion is *decreasing*.
- *Capital committed* to a project *up front,* prior to prototype development, is *increasing,* as is corporate risk.
- Market competition is greater than ever before, with a greater *demand* for *low-cost, high-quality* products.
- *Increasing technology* and stiff *market competition* require *inflexible time* and *money* commitments.
- Advanced technology requires more and more *specialized* and *integrated project personnel.*
- Specialization requires an even greater need for *product* and *project interface procedures,* across a number of organizational elements.
- Complex technologies make *cost overruns* common and *schedule slippage* more likely.

Thus, decreased time, increased costs, specialized personnel, and tight competitive markets force managers and project specialists to derive more efficient and systematic approaches to manage the dynamic and changing environment characteristic of high-tech projects and programs. No longer can such "mega-projects" be run by "seat-of-the-pants" management techniques. Nor can traditional management practices, originally established to generate and control repetitive, standard tasks, within a highly structured and predictable hierarchy, meet the needs of the politically sensitive and changing project environment. Management practices and principles must be tailored to accommodate the unique, dynamic, and diverse project needs while attaining better control of corporate resources. Project management tools and procedures are now more actively and openly under consideration to meet this challenge.

1.3 WHAT IS PROJECT MANAGEMENT?

1.3.1 Providing Definition and Direction

It never ceases to amaze our workshop participants when, after 45 minutes or so into the first session, we turn to them and ask, "Okay, what does 'project management' mean to you? Why are you here?" At times, with an audience average of 200–300 years of management and technical experience sitting before us, it is still amazing to see how eager most managers and technical personnel are to learn of these "magical tools" for tracking and controlling projects and programs. What is even more bewildering, however, is how many different perceptions there are of just what project management *really* is.

As advisors and consultants to both the commercial and military sectors, we have seen and heard a lot of what corporations and government agencies perceive project management to be. Unfortunately, the wide variety and disparity of commercial products, the markets, as well as business of government agencies that apply project management techniques to their operations, makes a clear-cut, generic definition of the "art" of managing projects a very difficult task.

As far back as 1917, Henri Fayol[1], Father of Management Science, identified the functions involved in the management process. These functions, frequently referred to in basic management courses and supervisory development seminars, include accountability and responsibility for *planning, organizing, directing, controlling,* and *staffing*. It includes a strong sense of command and *legal* authority over the contributory resources. Individuals, however, who find themselves in a visible and accountable project position, will discover the traditional vertical, superior–subordinate relationship *dwindling,* and the blending of company and project elements across divisions growing increasingly more important. As products and projects grow increasingly complex, cutting across political and geographic lines, managers and executives will require an approach that departs from a vertical emphasis and stresses the tools, procedures, and techniques for penetrating the formal "chain of command." Integration of efforts across the organization becomes the primary objective.

1.3.2 Project Management Functions

Traditionally, management practices have revolved around Fayol's five basic functions: planning, organizing, directing, controlling, and staffing. Although similarities can be drawn between traditional management and project management functions, *strong differences exist* that carry powerful implications for the individual in a project management role.

As indicated in Figure 1.3-1, project management functions begin with "integrative product or project planning." No longer is it efficient or effective to design first and determine manufacturing tooling, testing, or

Figure 1.3.1 Project management functions.

quality later. Under the leadership of the Project Manager, Manufacturing, Engineering, Marketing, Design, and Quality Control and all other project areas determine, define, and communicate project mission and objectives and agree on a plan for the project team to follow. As will be discussed in Chapter 3, one of the primary reasons why plans fail is that they have not first been developed by those who are responsible for their implementation. "Integrative product planning" involves the better use of resources. It encourages an organization and its people to interact not only in their traditional vertical manner but also diagonally and horizontally so that work flows more naturally and efficiently.

Plans, however, cannot stand alone. When developing today's complex systems and products, it is imperative for a project team not only to map out a course of action in moving from point *A* to point *B* but also to jointly establish and agree on checkpoints and standards. Such "hard milestones," or well-defined deliverables, communicate to the Project Manager, Functional Managers, and project team just what is expected and how well it has been achieved. Such communication enables the Project Manager to *control* the project.

Project control is not defined in the Machiavellian sense of the word. Nor is it characterized by unilateral decisions made by management decree. Effective project control involves *joint determination* of the tasks to be performed and the time required to accomplish them. *Participative control* is based on integrative project planning and requires key project team members to take meaningful and timely action when feedback indicates that corrective measures are necessary.

Above all, project management requires the recruitment, leadership, development, and commitment of people who truly make the project management phenomenon occur! As project efforts cut across traditional lines of authority, the ability to influence individuals, decisions, and events becomes less an issue of formal position in the organization and more one of interpersonal effectiveness and informal bases of power. The complexities and structure of today's high-technology organizations demand a more "people-oriented" approach to the functions of project management. No longer are participative approaches considered a "nicety" but a necessity as an increasing number of project leaders find themselves operating in complex organizational webs. Project goals are more likely to be accomplished when they are perceived to be in accordance with the technical, professional, and organizational goals of the managers and specialists who make up the project team.

1.4 TOWARD MORE FORMAL DEFINITIONS AND MANAGERIAL IMPLICATIONS

Project management, thus, is the mixture of people, systems, techniques, and technologies necessary to carry the project or program to successful completion within *time, budget,* and *performance constraints.* As organizational objectives with regard to market strategy, products, and/or services vary, so, too, do the more formal approaches and procedures of a project management system.

If we were to establish a more "formal" definition of project management, we would state the following:

Project management is the planning, organizing, directing, and controlling of company resources (i.e., money, materials, time, and people) for a relatively short-term objective. It is established to accomplish a set of specific goals and objectives by utilizing a fluid, systems approach to management by having functional personnel (the traditional line–staff hierarchy) assigned to a specific project (the horizontal hierarchy).

Traditional management approaches may, of course, be modified and applied to project ventures. However, the dynamic and unique qualities of projects do require innovative techniques that differ from the more traditional. These characteristics and implications of projects that warrant a "new" managerial approach include *a short-term objective, a systems perspective, life-cycle management,* and *the issues of power and authority.*

Short-Term Objective. Projects are run for a relatively short-term duration. Commercial engineering projects, for example, typically run a minimum of 6 months to 2 years, while some military endeavors, such as the development of a weaponry system, may run from 5 to 10 years. As Figure 1.4-1 indicates, project management planning, scheduling and controlling procedures were designed to keep project efforts on target. These strategies enable project managers to better manage and control company resources within a temporary and changing environment, and still address the constraints of time, cost and performance specifications. This requires the use of planning and control systems that quickly track and document the budgets, schedules, and technical specifications. Because of the dynamic nature of short-term project objectives, traditional management tools do not readily measure and provide feedback for corrective action.

It is important to balance the relationship between the triple constraints of time, money, and performance and the product quality. In today's highly competitive and rapidly changing marketplace, quality performance is an outgrowth of knowing what technical approaches can best meet engineering objectives, while possessing a firm understanding of what the customer perceives quality to be. No longer a nicety, but a necessity, customer responsiveness may actually be conceived of as the fourth dimension impacting project decisions and endeavors.

The project manager is in a vital position with regard to influencing and assuring the quality of a product. As a *liaison* across the project organization and a focal point for all life-cycle activities, the project manager can

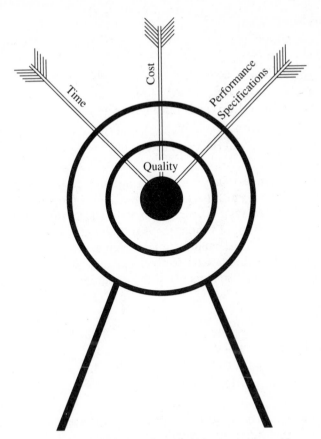

Figure 1.4-1 The mission of project management.

influence the choice of design alternatives and specifications to ensure the quality of the system.

The Systems Perspective—Interface Management. Projects, like systems, consist of a collection or assemblage of people, things, information, monies, and so on. To manage a project efficiently, it is best to manage "as if" one were managing a total system. Such a "systemic" approach to project management requires the break down and identification of each logical subsystem component (e.g., facilities, hardware, software, management approaches) into its own assemblage of people, things, information, or organization required to achieve the subobjective. And, as in most smoothly run systems, "synergistic integration" succeeds in making the entire system run in a way that surpasses the performance and capability of each individual subcomponent.

To run as a successfully operating system, projects require special management strategies that emphasize the following:

- The need for clearly defined objectives.
- The recognition that projects are organizations undergoing constant change.
- The need to clearly define and manage each subproject system.
- The importance of system control techniques such as feedback, widespread communication, and transmission of information.
- The need to determine and manage subproject interface points.
- The demand for "interface management."

Those who experience success in managing engineering projects discover the true sense of the term "interface management." Originally conceived as a technique to ensure system interface match, successful project engineers and managers now think of "interface management" to define not only the organizational, managerial, and technical components, but also the behaviors and activities required to effectively manage organizational relationships. Project interface management thus requires Project Managers to identify:

- The components or subsystem(s) to be managed on the project.
- The main subsystem interface points requiring management attention.
- The interpersonal techniques that need to be emphasized to manage these interactions effectively.

Life-Cycle Management. One major reason managers and project leaders resist the implementation of formal project management techniques is the complex, dynamic, and changing nature of the project system. Typically, systems go through a distinct life cycle in which a natural order of thought and action is followed. Projects, like systems, evolve through phases of development marked by a beginning, a buildup, a happening, and an end.

In each phase of the project life cycle, different levels and varieties of technical and managerial activities and procedures are required that generate their own complex set of problems and solutions. Projects typically are "born" when a sponsoring division within an organization identifies what may be either an external or internal customer need and accepts the challenge to accomplish the goal through project efforts. As planning and the execution of plans occurs, money, personnel, time, facilities, and other resources are committed to this challenge.

Each project life-cycle phase requires managerial techniques to meet the demands of the changing system or product. During the initial phase of the project, for example, innovative product champions identify the need for the project venture and establish a bare-boned project team. This project planning task force gathers information from across the organization. It determines the mission of the project, estimates the resource requirements, and "sells" the concept to higher management. As the project moves from concept to definition and the project team is firmly in place, specific project, technical, and administrative plans are jointly developed, while schedules are mapped out and contingencies for high-risk areas are identified. Once into product design and development, work breakdown structures indicate the work to be accomplished, as continual testing monitors performance and assures that quality standards are met. As the project approaches installation and eventual phaseout, the high-stress environment is loosened, while efforts concentrate on the reallocation of resources and the identification of new features.

Issues of Power and Authority. High-technology project efforts, with their complex and unstructured tasks, typically create working relationships that cut across divisional lines of authority. Even in what many companies today refer to as "projectized" or "project organizations," Project Managers often find themselves relying on other divisional or organizational personnel, such as graphics support, computer-assisted design (CAD), or engineering reliability, to help them accomplish project objectives. The Project Manager, as the focal point of project activities, is central to this "cross-pollination" and interactive organizational effort. Faced with *total responsibility* for project activities, *maximum accountability* for project success or failure, and *minimum authority* to direct project personnel, a great deal of effort is spent negotiating with and influencing others who can provide the technical, administrative, and marketing know-how needed for project completion. Project managers must rely on a variety of interpersonal, organizational, and political skills to develop the necessary support from functional specialists and minimize the conflict situations inherent in managing today's high-technology efforts.

1.5 WHY PROJECT MANAGEMENT CAN FAIL

The reasons for project management failure are as many and as diverse as the wide variety of activities, products, and services that come from high-technology organizations and efforts.

The process of planning, organizing, controlling, and directing company resources is practiced with minimal difficulty when members within an organization are focused on rather explicit, self-contained, and highly repetitive tasks. When work becomes more complex, however, and a high degree of specialization is involved, highly complicated interorganizational and interpersonal efforts are required to assure project completion and success.

High-technology systems and products share a variety of characteristics that illustrate the tremendous coordination challenges involved in managing project efforts:

- Increasing size and complexity.
- Specialization of skills (i.e., engineer, marketeer, technician, manager).
- Diversity of organizational and personal goals.
- Frequent technological and organizational changes.
- High degree of uncertainty.

Unfortunately, the traditional line–staff structure characteristic of most organizations has not demonstrated the flexibility required to develop and implement large-scale sophisticated systems. When the form or structure of an organization fails to follow the activities and functions of the organization, it is impossible to maintain the flexible utilization of resources needed to manage project resources properly over the duration of the life cycle. More often than not, complicated and dynamic project efforts are inappropriately forced into a structure or organization in which work tends to be highly repetitive and carefully prescribed. Incomplete or poor definition of the project activities and mission, compounded by the lack of involvement of all contributing organizational elements in the planning stage, are signs that project difficulties, reflected in schedule slippage or inferior product quality, will be experienced down the road. Some common reasons for project management failure are as follows:

1. Lack of a project focal point.
2. Poor choice of organizational form of structure.
3. Project efforts in the hands of one of the lead functional groups.
4. Inadequate involvement of team members.
5. Inadequate planning.
6. Lack of top management support or project administration efforts.
7. Too little authority in the hands of the project manager.
8. Poor choice of Project Manager.
9. Team not prepared for team efforts.
10. Poor project communication.
11. Lack of team blending.
12. Unclear project mission.
13. Objectives are not agreed on; end result is unclear.
14. Inability to estimate target dates.
15. No hard milestones; little project control.
16. Poor planning of project installation and termination.
17. Poor technical and user documentation.

The goal of project management is to provide integrative managerial techniques to complex and unique organizational ventures. These ventures are typically characterized by interdependent organizational efforts; integration of a variety of specialists; and cost, schedule, and resource constraints. With no one person, no project manager, serving as focal point of project efforts, coordination and control become extremely difficult, if not impossible. Although many government agencies and commercial corporations have long-used "quasi-project managers" who serve as "coordinators," "monitors," or "project liaisons," the lack of authority assigned to these positions undermines project efforts and contributes to frequent conflict and naturally, to project failure. Faced with their inability to take direct measures to convince project personnel to take action to expedite project efforts, project coordinators must quickly learn the art of negotiation and develop a tolerance for frustration.

1.6 THE PROJECT MANAGER'S ROLE: CONCLUSION

Achieving total project integration, especially in a high-technology matrix environment, is a prerequisite for effective project management. Total project integration is concerned with not only securing requisite resources in the proper and most efficient proportions but also achieving a unity of effort from a variety of disciplines. Unfortunately, both these concerns occur in an environment that is

- competitive,
- resource-limited, and,
- conflict-ridden.

In their search for information, Project Managers must cut across divisions and disciplines. As the focal point of project activities, it is critical to project success for the Project Manager to integrate this diverse input into the project. Through *integrated* efforts, Project Managers may gain support, develop more accurate predictions, and establish more effective control.

This book deals with the problems and issues confronted by Project Managers. It is based on actual client examples and experiences drawn from a variety of government and commercial endeavors. It is presented with the intent of helping project managers and personnel develop and maintain the skills necessary to become more efficient and effective in this highly integrative and demanding role.

REFERENCES

1. Fayol, H. *General and Industrial Management* (London: Sir Isaac Pitman and Sons, Ltd., 1949), p. 8.

BIBLIOGRAPHY

Adams, J. R., and Barndt, S. E. "Behavioral Implications of the Project Life Cycle," in D. I. Cleland and W. R. King (Eds.), *Project Management Handbook*. New York: Van Nostrand Reinhold, 1983.

Butler, A. "Project Management—Its Functions and Dysfunctions," in D. Cleland and W. King (Eds.), *Project Management Handbook*. New York: Van Nostrand Reinhold, 1983.

Carlisle, H. *Management: Concepts, Methods, and Applications*. Chicago: Science Research Associates, Inc., 1982.

Cleland, D., and King, W. *Systems Analysis and Project Management*, 2nd ed. New York: McGraw Hill, 1975.

Dill, D., and Pearson, A. "The Effectiveness of Project managers: Implications of a Political Model of Influence," *IEEE Transactions on Engineering Management*, Vol. EM-31, No. 3, August 1984, pp. 138–146.

Dinsmore, P. C. *Human Factors in Project Management*. New York: American Management Association, 1984.

Galbraith, J. K. *The New Industrial State*. New York: New American Library, 1968.

Hosley, "Innovative Approaches to Quality Assurance in Project Management," *Project Management Journal*, June 1985.

Kernzer, H., and Thamhain, H. *Project Management for Small and Medium Size Businesses*. New York: Van Nostrand Reinhold, 1984.

Morris, P. "Managing Project Interfaces—Key Points for Project Success," in D. Cleland and W. King (Eds.), *Project Management Handbook*. New York: Van Nostrand Reinhold, 1983.

Thamhain, H., and Wileman, D. "Leadership, Conflict, Program Management Effectiveness," *Sloan Management Review*, Fall 1977, pp. 69–89.

Case Study*

THE TACTICAL RADIO
TR-1000

On June 5, 1977, a meeting was held at National Radio between Donna Kay, Executive Vice President of Marketing, and Steve Davis, Executive Vice President of Engineering. The purpose of the meeting was to discuss the development of a new product for the U.S. Army. The requirements would include a very tight budget and a very short development schedule. The key to the success of the project would depend on the ability of Engineering to design a product while keeping costs at rock bottom.

Davis: "What's the scoop, Donna? I know you guys over in the Marketing group have been conducting extensive market research studies. Are you going to let us in on it now?"

Kay: "Look, Steve. We know National has always been the leader in tactical radios. Why, our market line is unsurpassed!"

Davis: "So, what's the bad news?"

Kay: "Well, most of these radios are in the upper price ranges. And now with the increase in foreign competition, we in Marketing and Sales have determined that it's up to you guys to give us a tactical radio that just wipes out the competition."

Davis: "What are you looking for, Donna?"

Kay: "Cost is of the essence now, Steve. We need a tactical radio that can really compete in both the middle and lower price tiers. Our studies indicate that within 2–3 years the tactical radio market will be ripe for the picking, and if we design a radio now with frequency hopping capability, that can sell for about $10,000, we can beat the heaviest foreign competition."

Davis: "That's not going to be easy, Donna. What you're asking for is sirloin on a chuck-steak budget. It takes time to get a high reliability and quality product up and running smoothly. And time is money!"

Kay: "That's right, Steve. But all signs indicate that National had better begin designing this low budget product NOW! if we want to maintain our standing in the defense marketplace."

Davis "I see what you mean, Donna. Okay, I'll get the appropriate departments and people involved as soon as possible. You know that the TR-100 radio has been a pretty reliable product. Of course, it does

*This case study was written for class discussion purposes only. It is not intended to reflect management practices or products of any particular company or organization.

not have the capability to frequency hop. But maybe we can save some time and money if we can incorporate some of its features into the new design.''

Kay: "Now you've got it, Steve. If this new product can earn as good a reputation as the TR-100, we're home free! Just remember, keep those schedules tight and costs down!''

Thus, what turned out to be the initial product description of what was later to be referred to as the TR-1000 Tactical Radio was actually defined by Marketing. As Donna Kay had pointed out, this new product was to provide National Radio with a strong competitive radio that would dominate the lower price market; a radio that would complement and not compete with existing National product lines. In addition, market research indicated that the greatest potential volume of sales for this product was by the U.S. Army. This would additionally require the radio to be built for quite a bit of abuse. In order to accomplish these objectives, it was imperative that the costs to design and manufacture the TR-1000 be kept to a bare-bone minimum.

Design Begins

It was early January 1978 when Larry Thayer, Director of Electrical Engineering, called his four Department Heads together. Larry had received a directive about the TR-1000 from his boss, Steve Davis, and how it was time to get the Electrical Engineering Department moving.

Bill Jones: "What's this about National going into the moderate price market, Larry?''

Larry Thayer: "Well, Bill that's precisely why I've called all you guys together. Marketing feels that the time is ripe for us to get into the moderate cost range radio production. They want us to turn out a high-quality tactical-frequency-hopping radio, modeled after the TR-100, for $10,000. In fact, everything you'll need to know—competition facts, research, financial data, everything—is in this description that Marketing just bucked to me.''

Craig Williams: "Tell us about it, Larry.''

Larry Thayer: "Well, Marketing tells us that what's needed here is a tough, reliable, tactical hopping, radio—one that's targeted toward the U.S. Army. It's got to be rugged—but most of all—inexpensive! In fact, the most important aspect of this product's description is cost. As I said earlier, it should cost about $10,000.''

Harry Petersen: "All the specs are right there in that document, Larry?''

Larry Thayer: "Right, Harry! Marketing has told us everything from options needed, to the level of performance expected. Looks like our job is just to grind this baby out!''

Craig Williams: "Well . . . okay. Who's coordinating this job, Larry?"

Larry Thayer: "I don't think that there's any reason not to follow our regular approach. We've got everything mapped out for us in this product document. Craig, why don't you take the information from Marketing and combine it with the TR-100 specs. See what you can come up with and give me a technical description and block diagram by next Friday."

Fred Neumann: "I don't know, Larry. Maybe we should put someone in charge of this thing. Try a different approach this time. If cost is of the essence, I would imagine that time is, too. Remember the scheduling problems and delays we had in the last job we worked on? Why, by the time the review team finished with it, we had a totally different product and had to scrap most of our original target dates and schedules. It was a mess in my area! If we follow the same channels and procedures for this one, Larry, we are bound to experience too many delays and problems. Maybe someone should just track the schedule . . . you know, track delays and slippage."

Larry Thayer: "Well, Fred, it's something to think about. What do the rest of you guys think?"

Harry Petersen: I've been in this business for the past 15 years and I certainly know how to handle this job. I don't want someone breathing down my staff's back. They do their jobs and they do it well. Anyway, didn't you say that everything was spelled out for us in that Marketing description? I don't really see what the problem is."

Larry Thayer: "Okay, then, it's settled. I appreciate your concern, Fred, but Harry's right. Since everything we need to know is right here, all we have to do is determine how we are going to reach the defined goals. We can all schedule around the final target date. What other problems except the usual delays could we possibly be up against?"

Off to a Flying Start

As Harry Petersen had predicted, product design began with few quirks. After all, it appeared to be a simple job. Electrical Engineering had only to take the initial product definition generated by Marketing and respond with a feasibility report referred to as the Phase I Authorization Document. Design engineers evaluated all elements involved in producing the radio and determined which specifications could actually be developed and which would have to be eliminated based on product goals. Since the TR-1000 began with such a powerful product description, decisions were initially made solely on the basis of the number one criterion: cost. And with the constraints of time and money on them, there was a limited number of ways that Electrical Engineering could design this radio. The design appeared so cut-and-dry that instead of a large scale implementation plan to steer the

Table I   Phase I Milestone Chart—Effective Start Date: January 1, 1978

Task	Mean Time (months)[a]	Uncertainty[b]	Engineers Required (personnel-months)
Electrical			
Electrical Design	6	2	24
Simulation	6	1	12
Construct Breadboard	6	1	12
Testing	1	$\frac{1}{4}$	2
Final Drawings	1	$\frac{1}{4}$	1
Antenna			
Antenna Design	6	2	6
Construction	2	1	4
Testing	1	$\frac{1}{4}$	2
Final Drawing	1	$\frac{1}{4}$	1
Mechanical			
Chassis Design	4	2	8
Construction	2	1	4
Final Drawings	1	$\frac{1}{4}$	1
Assembly			
Complete Radio Assembled	3	1	4
Quality Assurance			
Testing	1	$\frac{1}{2}$	2
Documentation	1	$\frac{1}{4}$	4
Turnover for Manufacturing	–	–	–

[a] To complete task.

[b] Standard deviation–months.

development of this product, there existed only some initial milestone dates that were mentioned in the initial marketing description as shown in Table I.

Essentially, at this point the TR-1000 was the product of Marketing and Electrical Engineering. In fact, all design decisions and target dates were made by the Electrical Engineering Department, as dictated by the initial Marketing document. Mechanical Engineering, Manufacturing, and Quality Assurance had no essential input, lacked necessary data, and were merely alluded to in the Electrical Engineering Phase I Authorization Milestone Schedule. Members of the Electrical Engineering design team were just not concerned with any other groups' input. Advice was sought only when a problem arose.

Problems Begin to Surface

In spite of the anticipated ease of initial design, unforeseen difficulties began to arise that caused slippage and increased costs. Overlapping efforts occurred because poor planning left several fuzzy responsibility areas that could not be resolved until half-way through the breadboard construction.

Ignoring the suggestion to incorporate the circuitry of the TR-100, the circuit engineers believed that a cost savings could be accomplished by using new programmable logic arrays (PLAs) as well as ICs. This concept, however, required the engineers to learn to use a new development system, increased the number of design alternatives, and led to much conflict and delay, as the circuit designers argued over which technology should be used in the different parts of the design. To make matters even worse, it became evident after the first weekly labor report was analyzed that Electrical Engineering was spending $60 per hour, when the estimated cost was based on an average cost of $40 per hour. Investigations into these costs indicated that while a personnel loading table had been developed, it had not been used to determine costs, and that Electrical Engineering was using its most experienced circuit designers to respond to the perceived technology challenge and maintain the tight design schedule.

Product development concepts continued to be controlled by Electrical Engineering. But problems arose as a result of the lack of interface planning and communication in the face of limited, if any, systems engineering. For example, it was discovered that the dimensions of the chassis needed to house the radio greatly exceeded the dimensions specified by Marketing to meet the requirements of the U.S. Army.

As the second design review neared, Larry Thayer could not deny the uneasy feelings he had concerning this project. Budget overruns, schedule delays, and poor systems engineering were symptoms of a much deeper problem. He now hoped that once communication between all interested parties—Engineering, Marketing, Manufacturing, particularly Quality Assurance and Control, and Systems Engineering—was opened, a more effective means by which this product could be developed would be implemented. It was obvious to Larry that some formal project organization was badly needed. But how? "More aggravation," thought Larry. "It will probably just bring more aggravation."

QUESTIONS

1. What were the problems facing the TR-1000 development team?
2. How do these problems compare with the reason for the failure of project management?
3. What are the general indications that the traditional organizational structure may not be adequate for managing this project?

CHAPTER 2

THE PROJECT ORGANIZATION

2.1 INTRODUCTION

Arthur Chandler, in *Strategy and Structure* [1] expressed a notion that truly captures the forces and requirements of organizing for project work. Chandler's "form follows function" hypothesis quite simply observed that organizational structures in many of our more successful corporations are driven not only by the nature and characteristics of the products and services for which they are known but also by the pressures, challenges, and changes in the marketplace surrounding them.

Lawrence and Lorsch's [2] findings substantiate Chandler's hypothesis. They found that businesses that were characterized by stable, structured, slow-moving, repetitive products, such as a public utility company, were best suited by structures that maintained an uncomplicated simple form. Specialty companies, on the other hand, such as those in the dynamic communications industry, were more loosely structured and served best by more sophisticated systems to control the flow of work.

Sound organizational structure is basic to effective project control. But just what makes an organization "sound" for project work? Are flat, decentralized organizations, characterized by lateral decision-making, truly more efficient than layered, centralized structures in which decision-making is vertical, and higher-level management has greater responsibility for cost and schedule control? Debate regarding this issue has persisted for decades. There is no perfect organization. Traditional organizations with their vertical reporting relationships have been criticized for their complex lines of communication, rigidity, and resistance to change; while the newer, flatter, decentralized organizations create wide spans of control that may frustrate managers and demotivate employees unable to receive adequate direction and feedback from overworked supervisors.

Table 2.1-1   Advantages of Effective Project Organizing

- Resource assignments parallel technical scope
- Clear delineation of authority and responsibility relationships
- Minimal duplication of efforts
- Faster customer response
- Productive and cohesive work teams
- Minimal conflict between team specialists
- Employee professional and personal needs are addressed
- Communication and feedback mechanisms clarify objectives and priorities
- Communication and feedback mechanisms lead to early problem identification

Whether the structure is layered or flat, effective organizing offers a project several advantages. It assures that all functions and resources needed by the project are efficiently assigned and eliminates costly duplication of work. By carefully delineating the levels of authority and specific areas of responsibility, organizations can avoid the costly dysfunctional effects of conflict and help to build a more cohesive and productive project team. Efficient organizing helps to solidify organizations and make them more responsive to project and customer needs. It allows them to concentrate on the outside, where the true competition lies. Table 2.1-1 outlines several advantages of effective project organizing.

2.2   MEETING THE DEMANDS OF PROJECT WORK

The distinguishing characteristics that render a project environment unresponsive to traditional policies and procedures, have tremendous implications for building an organization that facilitates the accomplishment of project objectives.

Project characteristics place great demands on the organization and on the people who are responsible for accomplishing project objectives. As indicated in Chapter 1, projects typically are dynamic, one-of-a-kind undertakings that have a specific beginning and end with a specific and well-defined accomplishment or goal. Projects exist for finite periods of time and move through a sequence of activities and events requiring constant surveillance and adjustment of people, materials, time, and money. The systemic, multidisciplinary, fluid, nature of projects calls for some form of horizontal coordination of talents and materials. This horizontal effort encourages a variety of technical opinions but, unfortunately, creates a tremendous potential for conflict. An organization that is engaged in project efforts must be structured to encourage and facilitate the accomplishment of project goals. This involves developing planning procedures, rewards, and

The Project Environment

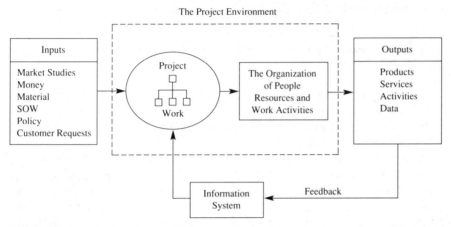

Figure 2.2-1 Project management process. Adapted from John Tuman. Jr., "Development and Implementation of Effective Project Management Information and Control Systems," in David I. Cleland and William R. King (Eds.), *Project Management Handbook*. New York: Van Nostrand Reinhold, 1983, pp. 495–529.

control mechanisms that parallel the nature of project work and direct human interaction toward goal accomplishment. Organizing for project work, therefore, must incorporate the following objectives:

- Provide a structure in which people may be encouraged and/or influenced to accomplish a particular task within the framework of the total project.
- Encourage collaborative efforts across the organization.
- Create a system that minimizes the duplication of effort and maximizes cost efficiency.

Figure 2.2-1 provides an overview of the project management process and the relationship of the organization to the project environment.

2.3 CREATING THE "RIGHT" MATCH

No one organizational structure will ever address all project needs. However, there are certain factors to consider when contemplating the "right" match. These factors include:

- The scope and/or technological magnitude of project efforts.
- Geographic dispersion of project team members.
- The nature of government regulations or requirements.

- Typical product lines or services.
- Outside environmental changes and future operations.
- Organizational culture, attitudes, and needs of the people who perform the work.

An organizational structure that would be effective for managing projects must be sensitive and responsive not only to the nature and characteristics of the technological products and services it generates but also to the social and professional relationship it creates.

Restructuring an organization to meet the challenges and demands of today's environment means more than analyzing and grouping technical tasks. It means assisting the informal "people organization" to adapt to the changes not only in the structure of the organization itself but also to the roles, relationships, and responsibilities that follow. Conflict frequently arises from the confusion of organizational change and adds to the existing pressures and difficulties of meeting typical time, cost, and performance constraints. A project organization should be established to provide sufficient channels of communication and properly integrated work, and to deal with the social and professional needs of its people. This is accomplished by creating a project environment that builds morale through the development of cohesive and productive teams.

Modifying the organization means much more than merely changing the reporting structure or regrouping functional tasks and assignments. People enter into organizations with a variety of expectations. These typically include the need for professional respect, technical growth, affiliation, job consistency, and security. Organizational change must include adapting systems, processes, and procedures to the social requirements of its primary resource—*people*. Nonhuman resources, such as materials and facilities, must be balanced against the needs and demands of the human resources to achieve a totally integrated project system, which encourages human interaction and minimizes the barriers to such interaction.

2.4 SYMPTOMS OF INADEQUATE ORGANIZATIONAL FORM

No organization can ever be perfect, and as Peter Drucker, management theorist and philosopher [3], aptly points out, at best an organizational structure will not cause trouble. But what are some symptoms of flaws in the project organization? Policy-makers may ask themselves the following questions when contemplating the need for reorganization:

- Is there a sense of product pride and ownership among the team members?
- Is too much attention typically given to one particular technical function, to the neglect of other technical components?

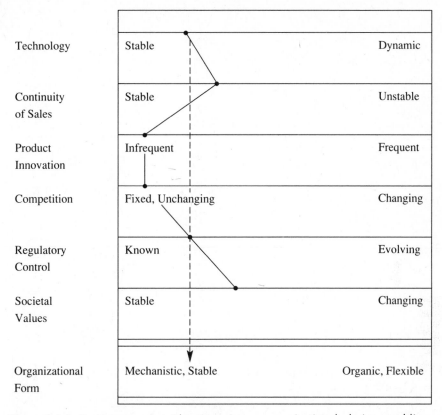

Figure 2.4-1 Contingency considerations in an organizational design—public or regulated industry. Adapted from Howard M. Carlisle, *Management: Concepts, Methods, and Applications*. Chicago: Science Research Associates, 1982.

- Does a great deal of finger-pointing exist across technical groups?
- Is slippage common, while customer responsiveness is negligible?
- Do project participants appear unsure of their responsibilities or of the mission or objective(s) of the project?
- Are projects experiencing considerable cost overruns as a result of duplication of effort or unclear delegation of responsibilities?
- Do project participants complain of a lack of job satisfaction, rewards, or recognition for project efforts?

Unfortunately, when symptoms of inadequate organizing appear, some companies typically respond by applying more time, money, or resources to the already weakened and inadequate project organization. If the problem truly is an inappropriately structured project organization, simply addressing the symptoms while ignoring the basic problem itself may leave the

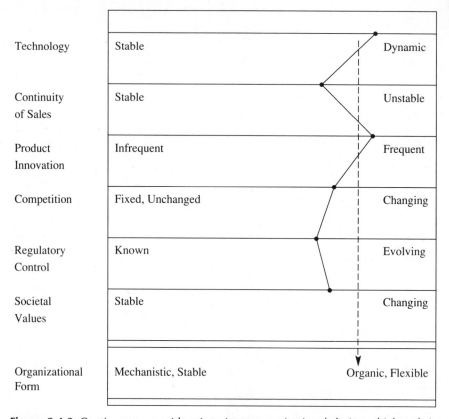

Figure 2.4-2 Contingency considerations in an organizational design—high-tech information company. Adapted from Howard M. Carlisle, *Management: Concepts, Methods, and Applications*. Chicago: Science Research Associates, 1982.

organization and its people frustrated and demoralized, as projects continue to slip and conflict continues to grow.

On the other hand, there are companies in which reorganization occurs at the first sign of confusion regarding, for example, specifications or customer interface. Such ''organizitis,'' as Drucker [3] refers to it, is a favored response of companies that have not clearly come to grips with certain fundamental changes in the size and complexity of business or changes in basic strategy and objectives.

A properly designed organization can be seen as a major contributor to project success. The most important factor to consider is that *the form of the organization chosen must parallel the dynamic and changing nature of the tasks and products it sets out to accomplish*. Figures 2.4-1 and 2.4-2 outline contingency considerations in contemplating project organization design.

2.5 BASIC REQUIREMENTS OF ORGANIZATIONAL FORMS

Project management involves the integration of company efforts within time, performance and budget constraints to accomplish a unique and dynamic organizational goal. A project organization must be designed to facilitate the work of project integrators. Minimally, the structure or form chosen must satisfy some basic requirements or specifications. These requirements include:

1. *Clarification of Roles and Relationships.* The organizational structure must indicate to each manager and specialist, alike, where that individual belongs in the organization, where to go for information or cooperation, and to whom one's problems can be escalated should such cooperation or information not be forthcoming.

2. *Efficiency and Economy.* The organizational structure should encourage teamwork, increase internal and informal communication, and minimize the need for elaborate and extensive control mechanisms. It should, therefore, minimize "input effort" and maximize company "output" to create an economical organization.

3. *Classification of Company Objectives, Goals, and Mission.* The organizational structure should clearly indicate the strategy and direction of the company and emphasize how performance leads to specific results. The more individuals understand the direction of the organization, the more efficient and goal-directed their performance. Structures that misdirect employees may actually encourage managers and specialists to cling to old methods and products, thereby reducing efficiency and profitability.

4. *Clarification of Individual and Organizational Tasks.* The organization should be structured so that all team members not only understand their tasks, but also know how those tasks fit into the specific project mission. Communication, therefore, must be facilitated by the organizational form.

5. *Encourage Decision-Making and a Rapid Response.* The organizational structure should strengthen reaction time and the decision-making process by minimizing channels or impediments that may block the efficiency of this proces. Structures may make it difficult for decisions to be made. Decisions need to be made quickly at the right level and converted rapidly into work and accomplishments.

6. *Provide a Sense of Stability and a High Degree of Adaptability.* Especially in the dynamic project environment, an organization must provide a project team with a sense of stability and a feeling of continuity. It must further be able to adapt to the new demands and market changes that occur around it.

7. *Encourage Innovation and Development.* For an organization to

survive, it must be structured to encourage the development of future leaders with innovative concepts for tomorrow's marketplace. The structure must encourage each employee to learn and develop maximally for the future, while presently contributing to here-and-now project objectives.

These seven basic requirements apply to any organizational structure determined to be appropriate for project work. We should now discuss several of the more popular and effective organizational structures that are available to facilitate project efforts. Each has its advantages and is best suited for a particular product situation. Many companies actually employ a combination of these forms in ways that best suit their culture, mission, and project objectives.

2.6 MODELS OF ORGANIZATIONAL FORMS

There are a variety of measures that may be used to effectively structure and organize complex project efforts. Since relations with customers, subcontractors, and general management are affected by the lines of authority and the reporting relationships that organizational structures create, it is wise for all project team members to be aware of the characteristic methods of organizing project work. Then they can fully comprehend their roles and expectations and adequately plan for the problems and difficulties that may lie ahead.

Because it is virtually impossible for us to describe the plethora of organizational structures that have been applied to a project management approach, three generic forms are discussed here:

- *The functional organization*—characterized by stratified levels of management, vertical lines of authority, and work that is partitioned according to specialties of disciplines (i.e., the "function"). The objective of a functional organization is to emphasize technical excellence.
- *The project organization*—characterized by pooled or "chunked" resources, temporarily banded together for a unique project effort and centralized under the authority of a Project Manager. The objective of a "pure" or "direct" project organization is command, control, and communication (C^3).
- *The matrix organization*—a mix or hybrid structure, incorporating both the functional and project approaches, with the hope of optimizing the strengths and minimizing the weaknesses of each. The objective of the matrix is, therefore, control *and* technical excellence.

Sections 2.7, 2.8, and 2.9 discuss each of these organizational forms in greater detail.

Organizing for a project management approach may thus assume a variety of forms, from the traditional functional grouping of work to the complex web of diagonal and horizontal matrix relatonships. Each form has certain *advantages* and *disadvantages* for project efforts; no one form is best suited for all project applications. In choosing the "best" approach, one must consider the size and geographic dispersion of the organization, along with the number, complexity, and duration of projects, customer interface requirements, job security, and management philosophy. Only through a detailed study of project requirements can implementation of the "right" structure be accomplished.

2.7 THE FUNCTIONAL ORGANIZATION

The functional organization, as depicted in Figures 2.7-1 and 2.7-2, is characterized by the classic division of labor and services and by vertical lines of authority and reporting relationships.

Organizing along functional lines is a common practice among companies in which Manufacturing is the predominant organization or "line of business." Company activities tend to be devoted to the assembly and production of standard products, and specialists are grouped according to the functions they contribute. Each "function" or discipline creates a pool of experts, where knowledge and experience can easily be shared. As part of their efforts to maintain more or less standard products, functional departments are involved in a large volume of highly structured and repetitive tasks. Because activities and tasks tend to be repetitive, uniform policies, procedures and standards can be applied readily.

At times, when a project effort appears to be uncomplicated or of relatively short duration, the manager or supervisor in charge of the "lead" or dominant functional area may be delegated the additional responsibility of managing the project. In the example of a tactical radio that appears to be clearly an electrical engineering function, the Manager of Electrical Engineering or Communications Systems, for example, may be asked to "pull the project effort together."

It should be obvious by now that to develop a "total" product or system, project efforts can rarely be fully contained within any single function. To develop a tactical radio, for example, the manager of Electrical Engineering is required to solicit the support of *all* functions (e.g., Mechanical engineering, industrial engineering, drafting and testing) to fulfill project specifications. The Functional Manager, wearing the two hats of Electrical Engineering and Project Manager, must cut across functional lines and obtain the required resources and commitment necessary to realize project efforts. This integration of project disciplines and activities is a difficult task,

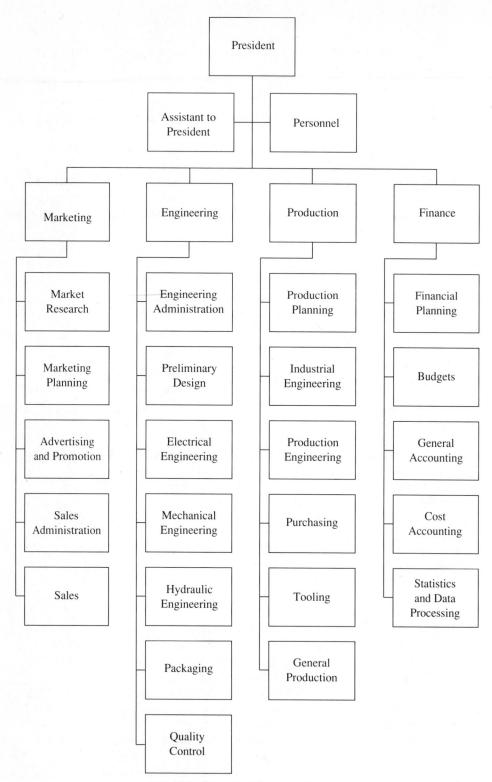

Figure 2.7-1 Functional organization.

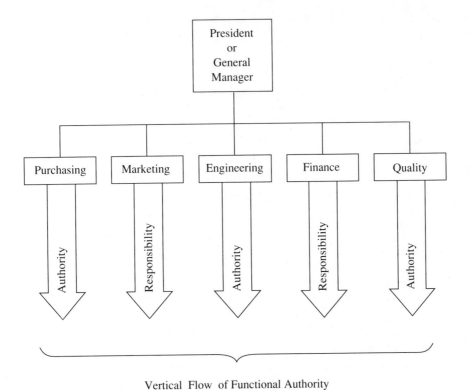

Figure 2.7-2 The vertical flow of authority and responsibility in the functional organization.

especially when it is in addition to one's everyday functional responsibilities. Conflict, moreover, is frequently created, as contributing departments vie for project power and control. Technical decisions may, unfortunately, be made to enhance the power position of each functional contributor, rather than for the good of the project. Decision-making becomes a slow and tedious process, as commitment or "buy-in" to what may be perceived to be a low-priority project, must be continually developed.

In the absence of a strong project focal point, communication between the functional organization and customer representatives is often weak and confused. Often, upper-level management serves in a customer relations capacity and funnels down project information through a very formal and layered chain of command. Response to customer needs, therefore, becomes a slow and tedious process. Schedules are often adversely affected, as the process required for the approval of decisions and changes is time-consuming.

Tables 2.7-1 and 2.7-2 outline the several advantages and disadvantages of the functional organization for meeting project objectives.

Table 2.7-1   Advantages of the Functional Organization

- Logical reflection of company or agency functions and disciplines
- Maintains power and prestige of the major organizational functions
- Allows for continuity in functional procedures and methodology (e.g., everyday administrative policies and technical procedures are easily retained)
- Retains control over personnel in the hands of the functional manager
- Provides clearly defined communication channels
- Follows the principles of occupational specialization
- Maintains a pool of specialists and talent
- Encourages technical and professional growth and advancement
- Develops technological expertise and future leadership

Table 2.7-2   Disadvantages of the Functional Organization

- Lack of a big picture, total project orientation
- No project focal point
- Less concerned with project objectives and goals than with functional activities and objectives
- Decisions tend to favor lead function or strongest group
- Lack of project emphasis with regard to schedule and cost
- Slow communication and decision-making processes
- Lack of flexibility and responsiveness to changing and dynamic project environment
- No customer focal point

2.8 THE PURE PROJECT ORGANIZATION

The Pure Project Organization, as shown in Figure 2.8-1, is often referred to as the "vertical," or "direct" project organization. It frequently emerges from the traditional or functional organization in response to high-priority project demands. It is, at times, likened to a task force organization, in which all players or team members report directly to an appointed Project Manager for the duration of the project life cycle. The Project Manager's position is clearly separated from the Functional Manager's, as the former maintains complete line authority over project activities and project team members.

Complete line authority over project efforts affords the Project Manager strong project controls and centralized lines of communication that lead to rapid reaction time and impressive customer responsiveness. Project personnel, moreover, can be retained on an exclusive, rather than on an "as-needed" basis. Project teams develop a strong sense of product

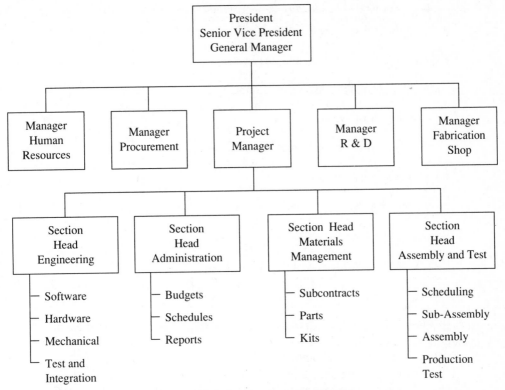

Figure 2.8-1 Pure project organization.

identification and ownership. Loyalty to project efforts is clear. Little doubt exists as to the nature of the project's mission or objectives.

Pure project organizations are especially common among long-term, technologically complicated, and unstructured projects, such as those characteristic of the aerospace industry. These large-scale and complicated multi-billion dollar projects can more easily absorb the cost of maintaining an organization that encourages companywide duplication of effort and less than efficient use of resources.

In fact, one major disadvantage of this form of project organization is the costly and inefficient use of project personnel. Project team members are dedicated to one project at a time, even though they are rarely needed full time over the entire life cycle. Project Managers tend to retain their star personnel long after the work is completed, preventing both their contribution to other projects and their professional development. Limited opportunities exist for technological exchange between personnel and projects, causing frequent complaints among team members over the lack of career continuity and opportunities for professional growth. With the elimination of

Table 2.8-1   Advantages of the Pure (Direct) Project Organization

• Centralized project planning emphasis

• Accountability clearly placed in one person

• Close coordination across project disciplines

• Maximum control over resources

• Commitment to project schedule, technical and cost goals

• Strong communication channels

• Early identification of potential problems

• Rapid reaction time

• Clearly defined customer focal point

a true "home base," project personnel tend to experience a great deal of insecurity and uncertainty, as shifts in company priorities or sudden project cancellations threaten job security.

Temporarily "projectized" organizations are another variety of the pure project approach. This organization consists of a project-oriented team, pulled temporarily from their functonal home base and led by a Project Manager with a specific charter that establishes some degree of project authority. This task force approach has been used by such corporations as Lockheed in its "skunkworks" method, and IBM with its organizational "chunking." Each of these techniques involves removing top-notch specialists from their functional home base for a period of time, primarily to generate a design plan [4].

The projectized task force is viewed quite favorably by those technical specialists who are actually involved in accomplishing the project tasks at hand. It is usually unhindered by organizational restrictions, policies, and procedures and carries many of the advantages of the full-blown pure project effort. However, since such task team efforts usually occur in response to an unexpected or unplanned project venture, the advantages they carry do not come without some major drawbacks. Since the nature of the projectized task force is temporary, the people who serve frequently are not relieved of their functional responsibilities. Moreover, as in the case of the pure or direct project organization, mobilization of personnel to a central location is difficult and tends to be quite costly.

Tables 2.8-1 and 2.8-2 outline the advantages and disadvantages of the pure project organization.

Table 2.8-2   Disadvantages of the Pure (Direct) Project Organization

- Lack of "big picture" companywide orientation

- Duplication of effort increases organizational costs

- Technical communication and development between projects is limited

- Conflict frequency occurs between projects over allocation of resources and personnel

- Difficult to share individuals across projects

- Project personnel limited to a single project effort

- Tendency to retain personnel longer than needed

- Difficult to assign individuals to new projects on project completion

- Lack of career development and opportunities for project personnel

2.9 THE MATRIX ORGANIZATION

It is unusual to find an organization that is strictly functional or purely project in form. More often, organizations represent a combination of the two. The matrix organization is an attempt at creating a hybrid structure that captures and blends the advantages of the functional approach with those of the pure project organization. Figure 2.9-1 illustrates a typical matrix structure.

The matrix is a complicated effort to maintain the consistency and excellence of technical groups, while addressing the need for an individual designated as a specific project focal point. Matrix Project Managers are delegated the total responsibility for engineering the project system across all contributing disciplines, departments, or sections. Functional Managers, on the other hand, retain full authority over technical specialists, ensuring that technical performance and standards are met, and overseeing the technical integrity of the project. Matrix Project Managers, therefore, are delegated total responsibility and accountability for overall project success or failure, while maintaining minimal formal authority over the specialists who perform the work. Typical authority and responsibility relationships between Project and Functional Managers are outlined in Table 2.9-1.

In spite of the often confusing and overlapping authority and responsibility relationships, the matrix has many advantages that make it desirable for project work. A flexible pool of personnel and resources is developed as

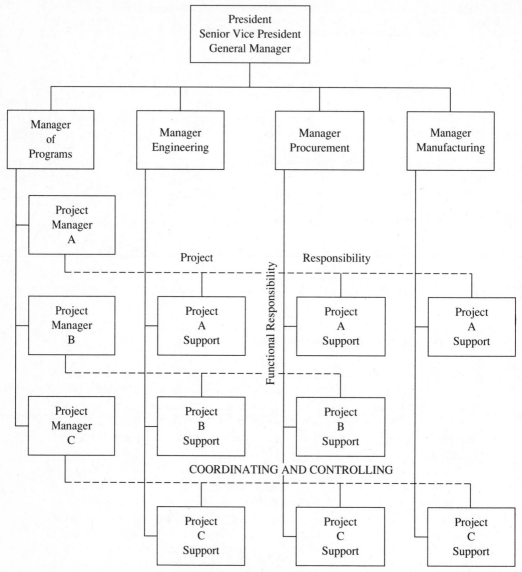

Figure 2.9-1 Matrix organization.

project personnel maintain their functional positions within their home units. Shifting talent demands, generated by the multiproject environment, can be more readily met as functional managers ensure that personnel are appropriately allocated. Costly duplication of effort is minimized, as knowledge and experience are readily shared across project efforts. Efficient use of human resources is realized through the use of project specialists on an as-needed basis. In fact, the shifting combinations of project personnel,

Table 2.9-1   Authority and Responsibility Relationships in a Matrix Organization: What, When, and How Much[a]

Project Manager	Functional Manager	Technical Specialist
Lead scoping activity for proposed projects	Participate in the development of project plans	Prepare detailed plans and schedules for design tasks consistent with overall project plan, including initial definition of support requirements

Represent project objectives to contributing functions in terms of:
  Cost
  Schedule
  Performance
	Participate in determination of project resource needs
	
	Define design workload
	
	Assign design personnel, consistent with project needs
		Execute design tasks
		
		Represent design engineering on project teams, in project reviews, for cost estimates, etc.

Lead project teams with regard to project definition, project schedules guidance, and direction	Maintain technical excellence of resources	Communicate
Request allocation of supporting resources with respect to technical needs	Recruit, train, and manage people in the organization	
Integrate and communicate project information	Assess quality of technical activities	
Track and assess progress against plan	Provide technical guidance and direction	
Serve as customer liaison	Communicate	
Resolve conflicts		
Communicate		

[a] Adapted from Cleland, D. I. and Kocaoglu, D. F. *Engineering Management*. New York: McGraw-Hill, 1981.

characteristic of the matrix, challenges any organization to realistically capture its form in any formal way.

Interactions in a matrix have often been likened to those of a marketplace. Negotiations concerning assignments, priorities, equipment, facilities, and people are constant. Matrix team members often complain of the continuous

Table 2.9-2   Advantages of the Matrix Organization

- Simultaneous operation of the functional organization and project efforts

- Advantages of the functional organization are maintained in an environment that is more "project responsive"

  —Functional managers maintains supervision over their own functional talent

  —Functional managers fill shifting talent needs of various projects

  —Functional managers promote desired technical standard

- Reservoir of talents maintained

- Flexible utilization of personnel across projects

- Opportunities for technological exchange

- Project team members maintain "home base"

- Project planning, scheduling, and cost control techniques are emphasized

- Project costs are minimized

- Provides a project–customer focal point

meetings, but it is through such meetings that the characteristic decentralized decision-making occurs. Group problem-solving and consensus-seeking activities replace individual and authority-based decisions, as commitment is obtained through a "buy-in," participative approach. Members must quickly grow to value and incorporate the diverse input, if project objectives are to be realized.

Project Managers working in a matrix environment quickly come to realize that if they are to truly manage their environment, they must win the support of their Functional Managers. This is not an easy task. Functions often resist being perceived of as "service" organizations to project efforts and will continually attempt to assert their autonomy and power. To make matters worse, Project Managers frequently are promoted to this position from the ranks of technical specialists and may be less experienced in management and negotiation skills, and therefore, less able to handle the conflict issues that lie ahead. If project integration skills are not quickly learned, the matrix frequently becomes a battleground for "we–they" issues.

The matrix organization has been highly criticized for its complex web of relationships and confused lines of authority. A concerted effort is needed to initially define problems and map out procedures. Through proper planning,

Table 2.9-3   Disadvantages of Matrix Organization

- Violates "one-boss" tradition

- Authority and responsibility lines are at times unclear

- Conflict frequently arises as a result of confused lines of authority and multiple reporting relationships

- Adjustment of personnel and managers is difficult and requires time

- Great demand for effective communication

- More effort and time are needed initially to define policies and procedures

- Matrix response time may be slow, especially on fast-moving projects

the scrutiny of present operations, and top management support, the matrix can provide a flexible means of adapting to an ever-changing, dynamic project environment.

Table 2.9-2 summarizes the advantages of this organizational form. Table 2.9-3 lists the disadvantages.

2.10 THE PROJECT OFFICE

We cannot leave our discussion of efficiently organizing for project work without a brief description of the Project Office.

The Project Office may be viewed as the "alter ego" of the Project Manager. If the Project Manager's primary responsibilities include planning, organizing, scheduling, and controlling project resources, the project office consists of those key individuals delegated with full-time responsibility for gathering and tracking the information necessary to support the project and ensure success. Project Office activities therefore include:

- Customer communication,
- Project policy administration,
- Project forecasting and planning,
- Project scheduling, tracking, and cost control,
- Reporting to management,
- Post-project evaluation.

The structure and size of the project organization (e.g., matrix vs. functional) as well as management philosophies and technical procedures (e.g., the use of a project management information system) affect the composition and size of the project office.

It is generally recommended that the number of persons assigned to the Project Office be kept at a minimum. This permits the Project Manager to devote energy mostly to the project effort, rather than to the administrative and supervisory duties required by a large staff. In addition, a leaner project office not only minimizes project personnel costs, but also requires each office member to work closely with functional personnel as a team in generating project data. Key personnel in the Project Office may include:

- *The project manager*—always considered the Manager of the Project Office. In the situation of a small project, the Project Office may consist only of the Project Manager.
- *A systems or project engineer*—responsible for the technical integrity of the project and for cost and schedule performance of all engineering phases of the project. This includes preparing technical-scope documents, assisting in the development of the engineering plan and budget, and developing the project design.
- *Project controller*—assigned to assist in the planning and control functions. This person may also be responsible for evaluating trends, predicting their effect on ultimate job completion, and advising the Project Manager of possible remedial measures.

2.11   PROJECT INTEGRATION AND INTERFACE

Determining the most efficient structure of an organization is only part of the task of efficiently organizing for project work. Effective project management begins with the integration of activities on a top-down, bottom-up, horizontal basis, across all major organizational boundaries or functions.

The project, as depicted in Figure 2.11-1, may be conceived of as a complex, multifaceted jigsaw puzzle. Its pieces are derived from such functional areas as Product Planning, Systems Engineering, Procurement, and so on, and must be fitted together neatly to create a total picture or product. Anyone who has spent any degree of time working on jigsaw puzzles will immediately realize the difficulty and time-consuming nature of the task. Organizationally, this fitting together of functional pieces is known as "horizontal integration" and is best achieved through a participative planning process, which is discussed in Chapter 3. In fact, *how* projects are planned, or the process of project planning, may prove to be more important to project completion and success than the actual plan itself.

Vertical integration (Figure 2.11-2) refers to companywide efforts to link corporate strategic plans with functional goals and daily project activities. *Successful* vertical integration can be achieved through management practices that encourage downward communication concerning the nature of the

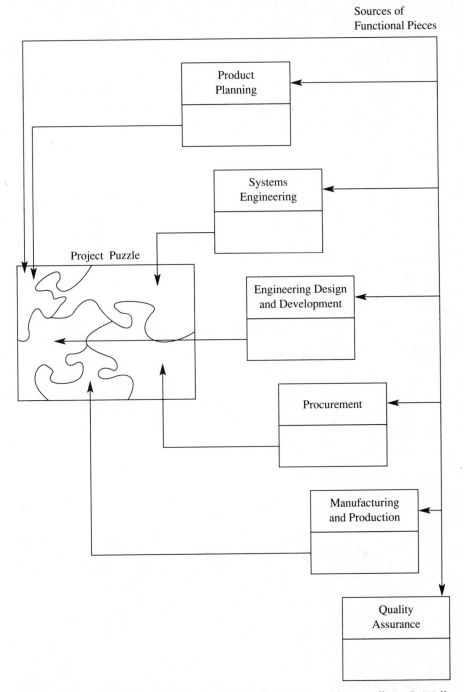

Figure 2.11-1 Horizontal organizational integration. Adapted from William C. Wall, "Integrated Management in Matrix Organization," *IEEE Transactions on Engineering Management,* Vol. EM-31, No. 1, February 1984.

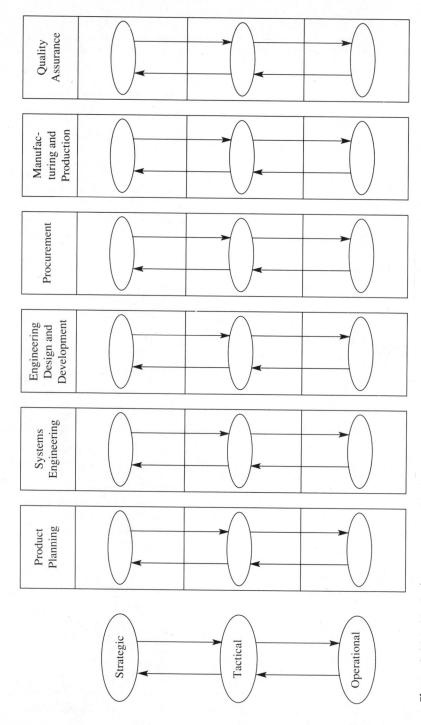

Figure 2.11-2 Vertical organizational integration is derived via integrated planning. Adapted from William C. Wall, "Integrated Management in Matrix Organization," *IEEE Transactions on Engineering Management,* Vol. EM-31, No. 1, February 1984.

project mission and goals and reinforce honest, upward feedback and reporting mechanisms. Upward communication must pinpoint not only project successes, but also potential pitfalls and problems. A "kill the bearer of bad news" philosophy only serves to bury problems that eventually lead to project slippage or technical failure.

Total project integration, therefore, is concerned not only with securing requisite resources in the proper proportions, but also with achieving a unity of individual and group effort. Unfortunately, the project environment is a highly competitive one, characterized by less than sufficient resources and designed so that conflict is inherent. By stressing consensus-seeking activities, responding positively to conflict, and learning to apply a variety of communication and negotiation skills, Project Managers may derive more meaningful explanations and accurate predictions and establish more effective and productive control.

2.12 THE QUESTION OF PROJECT INTERFACE AND AUTHORITY

Regardless of which structure or organizational form is eventually determined, conflict, slippage, and poor quality will eventually occur unless lines of responsibility and reporting relationships are clearly delineated. Matrix organizations have developed very poor reputations because of the confused multiple lines of authority and complicated communication channels. Actually, it is not the matrix itself that contributes most heavily to project slippage or technical failures, but the lack of a planned and orderly implementation of a structure that, indeed, has the potential of being confusing and chaotic. The most successful matrices are those that naturally arise from an environment that lends itself to multidisciplinary teams working together toward the accomplishment of a common goal. If lines of responsibilities and reporting relations are clearly determined and communication patterns formally established, the matrix can serve as a flexible and cost-efficient utilization of resources. Effective team planning, coupled with ongoing communication, is truly the key to any organization's success. Difficulties evolving from complicated project–organizational interfaces can be overcome if clearly defined lines of authority and areas of responsibility exist.

The Linear Responsibility Chart (LRC) shown in Figure 2.12-1, is one tool to help clarify and document project team responsibilities and task relationships. It is derived directly from the analysis of the project tasks, known as the *Work Breakdown Structure* (WBS) discussed in Chapter 3.

The Linear Responsibility Chart assures a Project Manager that every element in the WBS is properly assigned and accounted for. Like all project planning and scheduling tools, the LRC serves as an additional vehicle through which Functional Managers can buy in or commit themselves and their resources to the project.

Organization / (Who) — Project Breakdown Structure (What)		Project Management Office: Project Engineer Forward Subsystem	Project Management Office: Project Engineer Film	Project Management Office: Subcontract Administration	Engineering Department — Engineering Services Section: Drafting	Engineering Services Section: Model Shop	Engineering Services Section: Data Control	Mechanical Engineering Section: Structure	Mechanical Engineering Section: Shutter	Mechanical Engineering Section: Transport	Optics: Lens Design	Optics: Shop
Camera	P											
Forward subsystem		P										
Transport					S	S				P		
Structure					S	S		P				S
Lens					S				P		P	
Shutter					S							S
View-finder					S	S		S				

P — Prime Responsibility
S — Supporting Responsibility

Figure 2.12-1 Linear responsibility chart. Adapted from Harold Kerzner, *Project Management for Executives*. New York: Van Nostrand Reinhold, 1982.

The Communication Matrix (Figure 2.12-2) further delineates the lines of communication that are necessary to effectively and efficiently accomplish project objectives and avoid the confusion that results from faulty communication.

2.13 LOCATION OF THE PROJECT MANAGER

A question that frequently arises during the discussion of organizing for project work concerns the placement or level of Project Managers with regard to the organizational hierarchy and their power, prestige, and authority within the company. Naturally, the specific level at which a Project Manager is placed not only determines the Project Manager's project effectiveness, but also indicates the skills necessary to generate project commitment among team members.

A variety of factors serve to determine the Project Manager's reporting position. Size, scope, importance, and timeliness of the project are all critical elements. However, a basic principle or rule of thumb is that the Project Manager should report directly to that line executive or manager who has the power to resolve conflicts that may occur across project divisions. Thus, depending on the size and scope of the project, the Project Manager could feasibly report to a General Manager or to an Executive Vice President of a major corporate division. For very large-scale projects involving several divisions, the Project Manager would most likely come from the division with the primary contractual responsibility. In any case, one of the fundamental means of achieving high-performance project teams is to establish a clear position of authority in project situations. The need for clear and strong managerial authority for Project Managers is supported by the conclusions of a variety of empirical studies that report not only a positive relationship between high managerial authority and perceptions of project group effectiveness, but also a strong relationship between ambiguity in authority assignment and problem projects [5].

2.14 CONCLUSION: IMPLEMENTING PROJECT STRUCTURES

Unfortunately, the responsibility of managing a project is often delegated to persons whose position within the organization is quite low relative to the overall hierarchy of authority and power. This greatly limits a Project Manager's ability to rely on any "legal" or position authority when dealing with the conflicts that typically arise in a complex project environment. Project Managers' success is often contingent on their ability to negotiate with upper-level and functional management for the resources and support needed to achieve project objectives. Project Managers may increase their ability to deal with Functional Managers as well as upper-level management

Legend:
- ○ Daily
- ● Weekly
- □ Monthly
- ▲ As Needed
- △ Informal
- ■ Never

Reports to

	Internal							External (Customer) [a]						
	Project Manager	Project Office	Team Member	Department Manager	Functional Employees	Division Manager	Executive Management	Project Manager	Project Office	Team Member	Department Manager	Functional Employees	Division Manager	Executive Manager
Project Manager		○	●	△	▲	▲	●	○	○	■	■	■	■	△
Project Office	○		○	○	▲	▲	▲	○	○	△	△	■	■	△
Team Member	●	○		●	□	■	■	■	■	▲	▲	▲	■	■
Department Manager	▲	△	○		○	●	■	△	△	△	△	△	■	■
Functional Employees	▲	▲	○	○		■	■	▲	▲	▲	▲	▲	■	■
Division Managers	△	▲	▲	▲	▲		△	■	■	■	■	■	△	△
Executive Management	△	▲	▲	▲	▲	▲		△	△	▲	▲	■	△	△

[a] Does not include regularly scheduled interchange meetings

Figure 2.12-2 The communications matrix.

through a variety of techniques which develop commitment and increase project credibility. Such techniques, which are discussed in greater detail in subsequent chapters, include the following:

- Visibly incorporate others' opinions and ideas into project policies, plans, and procedures.
- Share credit and recognition with functional management as well as with technical team players.
- Anticipate and plan for project problems and pitfalls.
- Suggest solutions when pinpointing problems.
- Indicate clearly how your proposals or solutions will benefit not only the functional departments but the organization as a whole.
- Publicize and celebrate the accomplishment of project milestones.
- Make oneself visible and become well known among functional managers, top management, and team players.
- Build a reputation for conducting short, productive meetings.
- Adopt a participative style of management wherever appropriate.
- Learn to "lobby" and sell one's ideas on a one-to-one basis.
- Get to know the customer's wants, needs, and problems.
- Work on interpersonal skills to develop "charisma" and friendly relations.

REFERENCES

1. Chandler, A. D., Jr. *Strategy and Structure.* Cambridge, MA: MIT Press, 1962.
2. Lawrence, P. R., and Lorsch, J. W. "New Management Job: The Integrator," *Harvard Business Review,* Vol. 142, November–December, 1967.
3. Drucker, P. *Management: Tasks, Responsibilities and Practices.* New York: Harper and Row, 1974.
5. Peters, T. I., and Waterman, R. H., Jr. *In Search of Excellence.* New York: Warner Books, 1982.
6. Reeser, C. "Some Potential Human Problems of the Project Form of Organization," *Management Journal,* December 1968, pp. 549–467.

BIBLIOGRAPHY

Archibald, R. D. *Managing High Technology Programs and Projects.* New York: Wiley, 1976.

Dill, D., Pearson, A. W., and Jahri, M. "The Organizational Climate of Team Leaders in R & D: Demands, Support, and Constraints." *Proceedings of International Conference,* Copenhagen, Denmark, September 1982.

Dinsmore, P. *Human Factors in Project Management.* California: American Management Association, Publications Group, 1984.

Grinwell, S. K., and Apple, H. P. "When Two Bosses are Better Than One," *Machine Design,* January 1975, pp. 84–87.

Kerzner, H. *Project Management for Executives.* New York: Van Nostrand Reinhold, 1982.

Might, R., and Fischer, W. "The Role of Structural Factors in Determining Project Management Success." *IEEE Transactions on Engineering Management,* Vol. EM-32, No. 2, May 1985, pp. 71–77.

Rosenau, M., Jr. *Successful Project Management.* Belmont, California: Lifetime Learning Publications, 1981.

Case Study*

THE TACTICAL RADIO TR-1000 (Continued)

Phase II: Implementation

In August 1979 a second design review was held, which would have tremendous impact on the TR-1000's organization and procedures. Unlike the initial review, which involved primarily electrical design engineers and marketing representatives, the second review also included representatives from Mechanical Engineering, Antennas, Manufacturing, Systems Engineering, and Quality Assurance. The objective now was to have these experts from the various disciplines represented, "punch holes" in the design. And punch holes they did! Many modifications were made that would require Engineering to spend much more time in design modification and planning.

Larry Thayer again experienced that old uneasy feeling about the design modifications and the schedules backing the new TR-1000. A few days after the second design review, Ted approached his boss, Steve Davis, with his concerns.

Thayer: "Frankly, Steve, I'm a bit worried about the future of the TR-1000. At first glance it certainly appears that the product changes recommended at the second review meeting are going to blow whatever is left of our budget, not to mention our schedule! And now . . . with all those units involved, well, we've had our hands full just coordinating the efforts of the Electrical Engineering division."

Davis: "Look, Larry, I know you fellows have had your hands full. And considering all the constraints you've been working under, you've done a fine job. Top management has gone so far as to give their *full* support to the TR-1000 concept. This means more money and more people will be applied to the project."

Thayer: "Yeah. And more aggravation."

Davis: "Well, now that we've got the Antennas and Mechanical divisions involved, perhaps you may consider reorganization: come up with some sort of organization that will take the pressure off you, Ted, and bring this baby home."

Thayer: "That sounds good to me. I'll get everyone together sometime this week and kick a couple of ideas around with the guys from Electrical Engineering, Antennas, and Mechanical."

---

*This case study was written for class discussion purposes only. It is not intended to reflect management practices or products of any particular company or organization.

Developing an Organization

Although anxious to get the ball rolling again, Larry was not about to make the same mistake twice. Before *any* reorganization could occur, Larry knew that the design changes that had evolved from the second design review team had to be carefully scrutinized by the appropriate experts in the areas involved in the changes. Only then could tasks and activities be properly delineated and a suitable organizational form developed.

It was October when Larry stood before the Directors from Electrical Engineering, Antennas and Mechanical Engineering, and representatives from Manufacturing and Quality Assurance.

Thayer: "As you are undoubtedly aware, a notice from our Executive Vice President, Steve Davis, issued on September 5th, indicated full top management support of the TR-1000. We've already had additional financial and human resources applied to the project's budget. Steve also suggests that we study the design specs and come up with a more efficient organization that would suit the project's objectives and maybe help us to schedule the work a little more effectively. I'd like all of us to put our heads together and outline the organizational arrangements we see necessary."

Tony LaMotta, Director of Antennas: "Well, Larry I believe that our organization should reflect the major features of the product we are designing and eventually will attempt to build. In doing that, we'll be able to view our work in terms of larger but separable tasks that should somehow be integrated into a total program."

Susanne Plessey, Director of Mechanical Engineering: "Sure, that's what Project Management is all about. I think it's a good idea. We brainstorm the product design into very large separable tasks. Then once we get a picture of the organization, the coordination of all these activities should be in the hands of a Project Manager. This person, the Project Manager, would then be responsible for breaking all the main tasks into subtasks to which people are assigned until a complete program team evolves."

Thayer: "And we eliminate duplication of effort?"

Plessey: "Well, we're supposed to, anyway. Like I said, Project Managers and their teams integrate the various activities into a total program. I know different parts of the organization have been using this approach with increasing success."

Thayer: "Okay. Since you all seem to think it's worth the effort, I'll go along with the Project Management approach. Meanwhile, let's get all our experts together next Thursday. We'll brainstorm the 'major tasks,' as you put it, Susanne, and come up with some organization."

A Project Organization Evolves

The brainstorming sessions held by Electrical Engineering, Antennas, Mechanical Engineering, Marketing, Quality Assurance, Systems Engineering, and Manufacturing resulted in a reorganization of certain basic functions to meet project requirements and prevent duplication of effort. Mechanical Engineering design groups were combined to form a new function known as "physical design," and a supervisor was appointed to head this new function. Additional money and personnel were simultaneously supplied to the Electrical Engineering design area, although at this point of the project the contributions of Mechanical Engineering became more and more critical to the success of the product. The Organizational Chart, showing the Vice Presidential Areas, Directors, and Departments Heads involved with the TR-1000 Project, is shown in Table II.

But Problems Still Exist

Although reorganization had taken place rather smoothly, little was done to appoint a Project Manager to integrate the activities of the new organization. Schedules, cost data, and system design specifications continued to be maintained by each individual function. As may be expected, conflict arose over incongruent priorities and schedules. Communication between departments was still poor. The physical design team, for example, was heavily involved in making the radio appear to look smaller, while Marketing decided it wanted a waterproof product. No one cared what the other one was doing or when it was to be done. Yet, ironically, of major concern was the fact that a working prototype had not yet been produced.

And Now . . . Project Management

In November 1979 it was finally decided to *fully* implement the project management organization. The decision, of course, centered around management's concern regarding project deadlines and the need for teamwork among the design groups. It was further argued that if the Systems Engineering Department were fully involved this early in the project, fewer mistakes would be made, and the time between turnover to Manufacturing and the final production would be reduced.

The Project Manager. Clint Bronson was assigned the position of Project Manager of the TR-1000 because of expectional engineering skills. He was a young man in his late thirties, with Bachelor's and Master's degrees in Mechanical Engineering.

Clint had been with the company for 10 years. He began in the Mechanical Engineering division and rose quickly to Assistant Director of

50

Table II

Organizational Chart after Design Review Meeting

| Vice President | | Engineering
Steve Davis | | Marketing
Donna Kay | | Manufacturing
Andy Schwartz | |

Vice President
- Engineering — Steve Davis
- Marketing — Donna Kay
- Manufacturing — Andy Schwartz

Directors
- Electrical Engineering — Larry Thayer
- Drafting — Bob Thomas
- Testing — Ted Theobold
- Antennas — Lori Ready
- Mechanical Engineering — Susanne Plessey
- Quality Assurance — Sal Luca
- Factory Operations — Peter Sauer

Department Heads
- Hardware Development — Bill Jones
- Design — Craig Williams
- Simulation — Harry Peterson
- New Technology — Fred Neumann
- Physical Design — Sandra Kell
- Machine Shop — William Bennett
- Production — Bob White
- Materials Acquisition — John Close

that division. He developed the reputation of always meeting commitments relative to cost and scheduling. He was known as a very tough taskmaster, which at times led to conflicts, especially with functional group leaders in the project team. Although Clint considered cost and scheduling as most important, he also stressed meeting all specifications and required that all quality considerations be contained in the specifications.

As originally outlined in the Director and Department head October meeting, the Project Manager was responsible for the technical effort through all phases of the project and was also responsible for the control of project schedules and cost and, in a total sense, the profit of the project. As Project Manager, Clint would be responsible for the control of funds and their expenditures.

Clint believed that the first task of the TR-1000 team was to prepare the overall program schedule. In fact, as one of his first functions as Project Manager, he called a mass meeting of all persons involved in the design and production of the TR-1000. He announced:

> We've got a product to manufacture! We've all had our professional say in its design and development—but it does us no good if we can't get it out and on the market as quickly as possible. In order to accomplish this, we've got to get more of a handle on who does what and when it will be completed. In other words, I need a master schedule. I want a detailed account from each of you of every activity necessary for successful completion of the program. I would like you to make a list of every activity you know of which must be completed in your particular area, and what items or information you require to complete the activity. We'll then be able to properly track the progress of the program . . . but only if we list out every activity. I know that you've all been tracking the progress of your own functional activities up until now. Now let's put it all together and come up with a master plan. Please bring your list to Tuesday's meeting along with any questions you may have about the program.

The Revised Schedule

A hopefully final schedule, which was presented at Tuesday's meeting, is shown in Table III.

Building a Project Team. The work of the project office was carried out with the assistance of a very small staff and a few Assistant Project Managers (APMs). Clint chose his APMs carefully. They, in actuality, had to have the skills and technical abilities for coordinating the subsystem specifications and design. One APM was responsible for planning, cost, and schedule control and program evaluation and review technique (PERT) costing. In a sense, that APM performed the functions of a local comptroller and master

Table III Revised Schedule: 10/1/79

Task	Percent of Mean Personnel Months for Completion	Standard Deviation[a]	Preceding Task(s)
Electrical			
a. Electrical Design	–	Completed	–
b. Simulation	–	Completed	–
c. Breadboard	1	1/2	–
d. Evaluation & Testing	1	1/4	c
e. Drawings	1	1/4	d
Antennas			
h. Design	–	Completed	–
i. Construction	1	1/2	–
j. Testing & Evaluation	1	1/4	i
k. Drawings	1	1/4	j
Mechanical			
m. Design	–	Completed	–
n. Construction	–	Completed	–
o. Drawings	1/2	0	–
Assembly			
r. Complete Radio Built	3	1	e,k,o
s. Test & Evaluation & Quality Assurance	1	1/2	r
Documentation	1	1/4	s

scheduler. Another APM was chosen with a systems engineering background and was responsible for formulating the project's systems requirements and making sure that everything fit together, and that the entire radio worked after the antenna was added and the unit fit into the chassis. The APM for product integrity was responsible for developing and implementing a quality assurance program for the entire project.

Using information and input dates obtained from the functional managers, the APMs developed a detailed Work Breakdown Structure (WBS) as shown in Table IV. The WBS delineated detailed responsibilities across all project areas. As a result of the WBS, the project effort was subdivided into work packages, work units, and tasks for schedule, cost, and performance control. Job numbers were assigned at each level.

Table IV

Work Breakdown Structure (WBS)

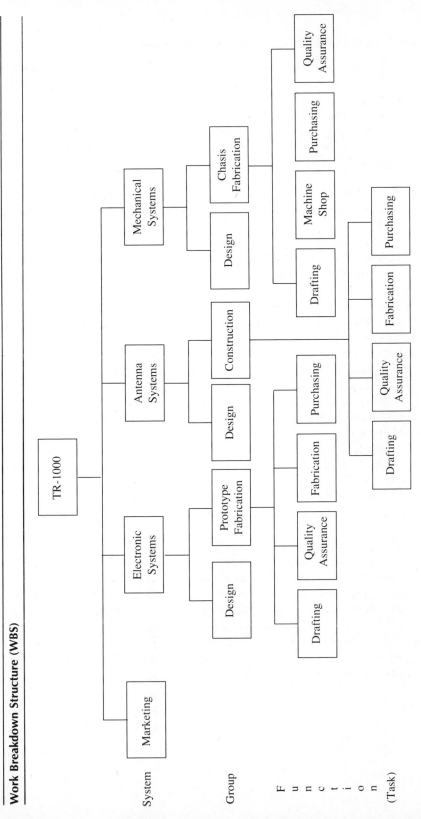

Implementing Project Plans via Communications

Because of their original objective, to manufacture a moderate cost-efficient tactical radio, National Radio management was quite concerned about keeping the project on track and within the agreed-on schedules. Clint Bronson was well aware of these concerns. One of the methods Clint used to ensure that project plans were implemented as agreed on and schedules met was to stress the importance of communication in the form of organizational meetings. As Clint pointed out:

> One of the things we decided to do as the key to ensuring that project plans became reality was to insure that functional shops always had the best possible grasp of what the task side of the house needed. If you want a schedule to *really* work, it looks to me to be vital that all the resource pieces of the organization have the broadest possible view of what the demands on them are expected to be.

In order to accomplish this objective, Clint instituted a variety of meetings that served to identify and resolve project planning and scheduling difficulties. For instance, every Monday and Friday he held "Roadblock Meetings" with Group Managers from Production, Systems Engineering, Quality, and Component Products. All the Department Heads and the Project Managers would talk about the problems in each division. This proved to be a mechanism whereby every part of the organization could question every other part and come to a better understanding of each other's contribution.

In addition to biweekly Roadblock Meetings, Clint held regular "Director Updates," which concentrated primarily on reviewing the schedule and discussing how major problems could have an impact on the schedule. He truly believed that these meetings lay at the core to the success of the project. As he frequently pointed out:

> *"Sure, you can map out schedules and they'll look great on paper. You can feed data to a computer and get a beauty of a PERT. But the most powerful tool that one has doing it, and it is really simple in a way, is openness. I try to create an open communication environment through my meetings. I encourage the managers and team members to tell me whatever they think I should know and, in turn, I have no secrets from them. If a manager makes certain resource demands, I need to know what's really going on in order to support him or her. The picture I get from these meetings is a really good one!"*

QUESTIONS

1. What were the factors the design and development team considered as they pondered the need for reorganization?

2. What tasks or responsibilities encompass the Project Manager's role in this new organization?

3. What changes in organization were made? How did these changes affect project efforts?

4. What problems continued to plague the development team? How did management attempt to resolve these problems?

CHAPTER 3

PLANNING THE PROJECT

3.1 INTRODUCTION

The importance of properly planning projects varying in complexity, size, and scope cannot be understated. Upper-level management and individual project team members alike contribute to this effort. The need to control corporate or government resources to accomplish specific strategic missions and goals over shorter and shorter time spans necessitates the development of more formal and detailed plans. Down from the executive to individual contributor level, planning facilitates the comprehension of complex problems and establishes policies, procedures, and programs that help to coordinate the effort to meet long-range goals.

Good plans are also essential for carrying out near-term project objectives. Without a proper understanding of the planning process, management will fail to execute the formal strategies necessary for achieving overall business and project goals.

The intent of Chapter 3, therefore, is to acquaint all active and prospective project personnel with the purposes of the project planning process as well as to teach specific techniques for planing the day-to-day operations, resource allotments, schedules, and budgets for the different life-cycle phases of a project.

3.2 TYPES OF ORGANIZATIONAL PLANS

Strategic planning, the capstone of all organizational planning, establishes corporate objectives. Strategic plans are long-range statements revealing a firm's purpose, goals, and policies for achieveing these goals. Strategic plans identify on a macro program level large-scale resource commitments that the firm will deploy to fulfill economic obligations to its customers, shareholders, employees, and the community at large. These plans are relatively

57

inflexible or irreversible, at least during a short time duration. However, they are characterized by a high degree of uncertainty.

Strategic plans answer such questions as, "What is our current business? What has it been in the past? What should it be in the future?" They generally project future business strategies based on technological forecasts of business activity as far off as 5–10 years.

Tactical plans relate to near-term time horizons that span a time interval from 6 months to 2 years. Performed by middle management, planning at this hierarchical level involves program or project elements more micro in nature. For instance, tactical plans detail company policies so that general guidelines for decision-making and individual action can be reasonably coordinated among projects and made known to all project personnel.

Tactical plans are generally more flexible and less environmentally influenced. There is less uncertainty or risk associated with them. Nevertheless, tactical planning is futuristic, as it involves determining the contribution a program or project will make to corporate long-term strategic plans.

Operational planning is the process of determining how specific tasks can best be accomplished on-time given the available resources. Although separated fiscally from the higher-level processes of strategic and tactical planning, a firm's operational budgets and plans are guided by the shared philosophies of these higher-level plans.

Operational planning is performed at the organization or project level by functional management and answers such questions as "What do we need to do? When do we need it? Can we do it? How long can we take? How do we monitor progress towards achieving these commitments?"

Typically, operational planning is determined by the calendar. Plans varying in duration from one week to one year are revised continuously at frequent periodic intervals. This type of planning characterizes such high-tech industries as telecommunication research and development, integrated component systems and device design, and software applications programming where a high degree of technological uncertainty and rapid change exists. Operational program or project plans focus on product development or process engineering. Corporate, integrated business units also produce an operational plan according to established, calendar-determined sequences of events [1]. The success of these industries depends on full integration of each planning level.

To summarize, Figure 3.2-1 contrasts the characteristics of each of these planning efforts and identifies who in the corporate hierarchy is responsible for the plan's execution.

3.3 THE CONSEQUENCES OF POOR PLANNING

Good strategic and tactical planning identifies the efforts managers at the project level must make to move the corporation toward realizing its future

Type of Plan / Characteristics	Strategic	Tactical	Operational
Hierarchical Level	Top Management •Chief Executive Officer •President •Vice Presidents •General Managers •Division Heads	Middle Management •Functional Managers •Product Line Managers •Department Heads	Lower Management •Unit Managers •Supervisors
Timeframe	Long Term, Time Horizon (generally, 5–10 years)	Intermediate, Near-Term Time Horizon (6 Months - 2 Years)	Short-Term, Time Horizon (1 Week - 1 Year)
Content	Vision of the Firm •Corporate Philosophy •Mission Identification •Strategic Business Units Budget Consolidation of Strategic and Operational Funds Allocation of All Resources	Formulation of •Corporate Performance Objectives •R&D Strategy •Planning Guidelines •Broad Action Programs Budgeting at the Business Level Deployment of All Resources	Work Authorization •Monitoring of Functional Level Cost and Control System Accountability for Performance Standards of Program Activities and Events
Approach	Formal, Inflexible, Irreversible	Formal, Less Rigid, Reversible	Formal, Flexible, Subject to Constant Updating

Figure 3.2-1 Characteristics of three types of organizational plan.

goals and improving its entrepreneurial performance. Proper operational planning documents the work of project participants and all commitments required of other key individuals who interface on specific project tasks. A good operational plan identifies not only required resources (human, capital, and equipment) needed to staff a project but also milestone events, and development schedules, as well as managerial accountability and authority. Finally, a good plan provides for measuring the results of the planning process itself as well as the work performed through the development of a *test* plan. When good plans result in successful execution of project objectives, the strategies that were initiated to achieve these efforts should be documented for future reference and adaptation by others.

Some of the consequences of poor operational planning are unsuccessful project initiation, disillusionment, and chaos at the project level. For example, if each functional department has developed its own planning documentation, a newly assigned Project Manager may discover that no effort was made to integrate the plans across corporate lines of business, vice presidential areas, or functional divisions. Typically, project objectives are not understood, resource allocations inadequate, schedules slipped, and budgets overspent because project requirements were not defined early in the planning process. Furthermore, failure to utilize systematic decision-making over the life of the project to update and monitor any deviations from a baselined plan results in uncertainty, discontinuity, and crisis, seat-of-the-pants-style conflict management. In short, progress is impeded, which results in the product being delivered late or released into an unresponsive market, and these discontinuities succeed in undermining the authority of lower-level management.

3.4 WHY PLANS FAIL

No matter how hard upper-level and project managers try, planning is not perfect and sometimes plans do fail. There are several reasons why even the most systematic efforts can go awry:

- Corporate goals and business strategies are not understood at the lower organizational levels.
- Strategic or long-range planning objectives are not redefined at timely and regular intervals when technology or external forecasts change.
- Planning is performed by a planning group without direct input from functional or line managers and other lead project personnel.
- Top management neglects to establish organizational structures designed to fit specific projects or direct the information flow requirements of the project.
- Planning stops short at the master schedule level; those who perform the detailed planning and control work are not consulted until the project is well under way.
- Plans are based on insufficient data.
- Financial estimates are poor.
- Not enough time is allowed for proper estimating.
- Estimates are best guesses not based on either standards, experiential, or historical data gathered on similar projects.
- Resource allocation is inadequate and corporate commitment for funding new projects short-sighted or insufficient for developing new technologies.

- No one knows the project's initial staffing levels or overall require-ments; typically, qualified personnel are available only part-time or simply are unavailable when needed.
- No one has defined the major work activities and tasks that must be performed, nor has anyone scheduled their associated milestone dates.
- End item specifications are nebulous and project deliverables not outlined in contractual agreements.
- Project personnel come and go constantly, and little regard is paid to the resolution of resulting schedule and resource conflicts.

3.5 PROJECT PLANNING TECHNIQUES

A project's success depends on *planning, monitoring,* and *tracking*—these are the techniques of project management. Refining these techniques produces optimal results: meeting predetermined specifications through the best use of available resources, in the least time and at the least cost.

Project planning must take into account the following concepts:

- Project objectives and strategies.
- An evolutionary life cycle of project planning.
- Preparation of a Project Summary Plan.

Thus, to be effective at determining what needs to be done, by whom, and by when, and at fulfilling one's assigned responsibilities, project planning must include:

- Overall project specifications and statement of work, including techni-cal performance, physical, and quality requirements, as well as target completion dates and cost objectives.
- Strategies for accomplishing these objectives. These strategies require the use of a *Work Breakdown Structure* (WBS) to delineate each of the tasks that need to be performed and the development of contingency plans for high-risk portions of the project.
- Forecasts of all project completion dates and a statement indicating the degree of reliability of these estimates.
- An organizational structure and design for the project accompanied by a detailed description of the authority, responsibility, and corresponding duties of all personnel—managerial, professional, and support staff.
- A schedule and budget for accomplishing the work activities that cover every aspect of the project, including contingencies, meeting and report due dates.
- Policy statements for making decisions that balance product quality and technical standards against the cost for completing by specified dates.

• Standards defining the criteria for acceptable and unacceptable individual and group performance.

The purpose of planning a project is to meet the *specifications* set forth for the product in a timely, cost-effective manner.

3.5.1 Project Specifications

The project specifications represent a statement, by the customer, of what that customer wants the product to do. In the *marketplace,* there is no single customer, and a good marketing person must determine what the generic customer requires and differentiate between those requirements and the customer's "wish list."

The specifications enumerate the product's performance and quality criteria. For example, it is not uncommon for specifications to state that (1) a transmitter be able to transmit at 347 MHz and (2) spurious signals above 348 MHz and below 346 MHz be 30 dB below the 347 MHz signal.

In addition, the customer can specifiy how the tests should be performed. Temperature, humidity, and shock specifications should also be given. In a production contract, the specifications also include a statement regarding the number of units, 48 of 50 or some similar number, that must pass the performance tests. The mean time between failure (MTBF) is another test that requires the product to be operated for a given time, 5000 hours or so, without failure.

During the planning phase of a project, the Project Manager (and/or Marketing Specialists) should discuss the specifications with the customer(s) to insure that the product is developed to meet a complete set of realistic specifications (see Section 3.6.1).

3.5.2 Project Objectives and Contingency Planning

Understanding project objectives optimizes the decision-making process of project planning. Through this early stage of the project life cycle, the manager is often called on to choose among alternative courses of action.

Decisions to develop a leading-edge product that is profitable and at the same time satisfies all the project's technical requirements are made more readily, when a reasonable analysis is carried out beforehand and documented in a preliminary project plan.

Planning for allowances or contingencies, on the other hand, is performed not only at the outset of the project but also throughout all subsequent life-cycle phases. Contingency planning is dependent on monitoring critical external factors that signal the need for developing alternative plans.

The following is a list of contingencies the project manager must plan for:

- Errors in estimations of the money, time, and personnel needed to complete the project.
- Design changes in the product that necessitate additional budget for fabrication and final production.
- Omissions from initial estimates to control for changes in scope, specifications, and estimated costs.
- Management reserves for expediting the work.
- Variations in the averages used to estimate resources required for project completion.

The *rule of thumb* for estimating reserves is 10%. Yet, because the level of risk and uncertainty is greatest during the preliminary start-up, design, and development phases of the technical project, these estimates may need to be as large as 20%. The best forecasts are obtained from analyzing past performances of similar R&D endeavors.

3.6 PLANNING OVER THE PROJECT LIFE CYCLE

To achieve their desired outcomes, all projects go through a specific life-cycle process beginning, first, with a formulation or *proposal* phase that includes technical definition and initial design, final design and development, and prototype system integration and test. On completion of these preliminary stages of their evolutionary life span, projects progress into the more mechanistic cycles of manufacture, production, installation, and field testing. Here, emphasis shifts from the more risky modes of project performance estimation to product control through monitoring how well a fabricated device operates within tolerance limits. The final, and often most difficult, phases for the high-technology research and development companies to accomplish are the turnover and project divestment or close-out phases. There is a tremendous tendency among high-achieving development engineers to "gold-plate" or "perfect their product" designs to excess of the specifications required, thereby failing to release them into the marketplace when customers most want them.

In this context, it is important to once again remember the *triple constraint: timeliness, specifications,* and *resources.* If a market window must be met, increasing resources by even 50% will rarely decrease profits by more that 5%. However, even a small delay can reduce profits significantly. It is often preferable to reduce the specifications by eliminating features than to be late.

Figure 3.6-1 illustrates the overlapping relationships of the many different life-cycle phases of the high-technology project. Each of the many phases is described subsequently in more detail.

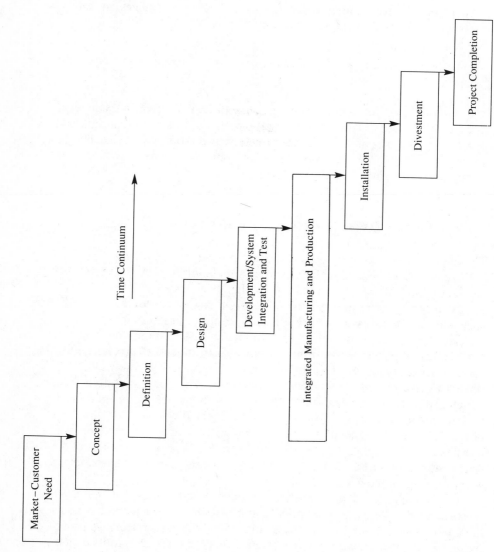

Figure 3.6-1 The life-cycle phases of a high-tech project.

3.6.1 The Proposal: Concept and Definition Phases

Typically, the proposal or preliminary planning phase is carried out by highly skilled groups of creative systems and design engineers who have been directed by upper management to establish the feasibility for a new technology, product, or service. In all cases the germs of product ideas defined in the early customer–market phase (see Figure 3.6-2) has set off this evolutionary process.

Management has three primary responsibilities during the concept and definition phases:

1. Timely solution of all business problems.
2. The allocation of adequate resources to staff the project start-up organizations.
3. Preparation of all preliminary project documentation (i.e., preliminary functional requirements, required quality standards, estimations of the tentative project duration, and cost objectives to the proposal team).

During the *concept* phase (see Figure 3.6-3) (1) upper management evaluates the economic markets and the feasibility of developing the proposed technology or systems in terms of corporate strategy, competitive advantage, risk, and an identified market need and (2) the proposal team prepares a preliminary design. To do this, the proposal team members need to interact with their customer, that is, whatever group furnished the specifications, to

CUSTOMER – MARKET

OUTPUTS
- Determined Market Need
- Preliminary Product
  Requirements Established
  Based on Commercial Analysis

PERFORMANCE STANDARDS
- Overall Business Mission
  and Strategy
- Defined Corporate Technical
  Strategy and Function of R&D
  Within Corporation

Figure 3.6-2 Customer—market phase.

```
┌─────────────────────────┐
│                         │
│        CONCEPT          │
│                         │
└─────────────────────────┘
                         │
                         ▼
```

OUTPUTS
- Engineering Design Concept
- System Feasibility Analysis (Prospectus)
- Initial Data as to:
  - Cost
  - Duration
  - Specifications
  - Development Risk

PERFORMANCE STANDARDS
- Approved Business Plan
- Proven Technical,Enviromental,
  and Economic Feasibility
- Identified Technologies in All Key Areas

Figure 3.6-3 Concept phase.

"round them out." This is an iterative process that is extremely important in proper planning of a project.

The "customer" usually specifies a system in an incomplete manner identifying only the *key* features needed. There are, however, other features: technical, quality, and human, that were not thought of by the customer. During the concept phases, the proposal team attempts to define these features and gain approval for them by the customer. This iterative process leads to a much better product, one that is delivered in a timely manner, within cost. Unfortunately, if this iterative procedure is omitted, there will be numerous changes to the specifications during the life of the project, causing delays, increased cost, and reduced morale among all project team members. To *minimize aggravation* and *maximize satisfaction,* we strongly recommend this first step in the life cycle.

In the *definition* phase (see Figure 3.6-4), specifics are added describing as accurately as possible the technical and economic requirements to develop the technology or proposed system. This is accomplished in an iterative manner between the proposal team and upper management.

To do this, the proposal team *partitions* the project into functional tasks. Then each task is partitioned, until no further partitioning is possible. This technique, called the *Work Breakdown Structure* (WBS), is discussed in detail in Section 3.8. Management uses the partitioned tasks to allocate

```
┌─────────────────────────┐
│                         │
│     DEFINITION          │
│                         │
└─────────────────────────┘
                         │
                         ▼
```

OUTPUTS
- Project Summary Plan
- Work Breakdown Structure
- Risk and Other Areas of
  Uncertainty Identified
- Master Schedule and Milestone
  Events Established
- System or Functional Requirements
  Analysis Completed
- Final Product Performance Requirements
  and Test–Qualification Requirements Prepared
- Identification Human and Nonhuman Resources
- Initial Project Documents:
  - Policies
  - Procedures
  - Budgets
  - Position Description
- Interface Points Determined for Each Major Task

PERFORMANCE STANDARDS
- Approved Project Plan
- Approved System and Functional Requirements
- Approved Product Performance and
  Test–Qualification Requirements

Figure 3.6-4 Definition phase.

personnel and other needed resources as well as determine the duration of the project.

Documentation. The primary planning documentation produced during these phases are the product or system prospectus and the R&D proposal, followed by a summary plan. The purpose of both the product or system prospectus and the R&D proposal is to obtain project funding and approval.

In the case of a regulated, public utility a prospectus to develop a system that supports its existing business would briefly describe the new system and the operations it impacts. The document would also include estimates for required resources, capital, other expenses, personnel, and computing requirements, as well as a high-level development plan or work program outlining proposed features and their development sequence, schedule, and intended release dates.

An R&D proposal, similarly, would describe the status of the new technology under investigation within and outside the company, including competition, and cite expected benefits to the corporation or government due to successful technical and commercial development. In other words, the proposal would attempt to determine what the payoff is for investing in this technology; what revenue and profits are anticipated; and what market share, cost reduction, quality improvement, energy savings, or compliance with government regulations are expected. The proposal may be the response to a solicitation from the government or an external, or internal customer. In such cases the proposal contains a description of the technical personnel and facilities available, various approaches considered, and the finally agreed-on approach for satisfying the specifications. A quantitative comparison of the approaches, including *risk,* is required. Duration, cost, and schedules, as well as management control techniques should be presented. These and other components of the Concept and Definition Phases are described in Figures 3.6-3 and 3.6-4. Further explanation of these outputs is covered in Sections 3.7 and 3.8 of this chapter.

3.6.2 Design and Development Phases

The previous phases produced the plan, which included the organization, and the schedules required to satisfy the detailed specifications and test procedures. Firm control of the system's technical definition is essential during both of these phases to ensure project success. The standard, functional requirements should primarily represent the customer's needs, constrained by what is technically feasible and economical to develop. In addition, unless these requirements are very carefully communicated to all members of the project development team, the system designed often ends up technically inadequate, user-unfriendly, late, and grossly over budget. This is particularly true of very large and complex software development projects where programmers sometimes tend to rush through requirements planning and immediately start writing code. (see Section 3.6.6).

The intent of the design and development phases is to:

- Produce working and tested models of the system using the specification and preliminary design defined in the previous two project planning phases.
- Determine the inadequacies of any previously conducted analyses.

In the *design* phase, the project team simulates all aspects of the system or product and then codes and/or constructs it. The resulting system should meet operational specifications with the exception of form, physical require-

OUTPUTS
- Detailed System, Component, and Module Design
- Software Simulation
- Hardware Simulation
- Breadboard Construction
- Completed Unit and Functional Testing
- Preparation of All Initial Drawings, Schematics, and Circuit Descriptions

PERFORMANCE STANDARDS
- Aproved Revisions To Project Plan
- Successful Completion All Unit and Functional Testing

Figure 3.6-5 Design phase.

ment, and final testing. In the *development* phase, the project team codes and/or constructs the final system and installs it in the chassis. This system should look and react like the final productized unit.

The Project Manager, therefore, is responsible for (1) reviewing with upper management, and confirming, the decision to continue the design effort; (2) developing the prototype system; (3) producing test plans and ensuring that test results meet specifications; and (4) developing procedures for installing the system at a qualification site. Project Managers should note that it is most advantageous to stop development efforts that fail to meet technical specification or appear to lead to a product cost greatly exceeding that needed to produce a profit. This is a strategic decision usually made by upper management on recommendation by the Project Manager. Also, overtime can be spent to greatest advantage during this period. By expediting the work when needed, to ensure timely project completion, maximum product profit can be obtained. Figures 3.6-5 and 3.6-6 detail the output and performance standards of these two phases.

3.6.3 Integrated Manufacturing and Production Phase

The purpose of this phase (Figure 3.6-7) is to transfer technology from the R&D laboratory to production. Planning this transaction is perhaps the most difficult in the soon-to-be-launched product's own life cycle because in large

```
┌─────────────────────────┐
│   DEVELOPMENT–SYSTEM     │
│     INTEGRATION          │
│      AND TEST            │
└─────────────────────────┘
                          │
                          ▼
```

OUTPUTS
- Prototype, Including Physical Design and Testing
- Proved-in Design of All Features, Structures, and Interfaces
- Final Drawings And Documentation:
  - Schematics
  - Circuit Packs
  - Equipment
- User Training Plan

PERFORMANCE STANDARDS
- Established Performance Specifications
- Scheduled Milestones and Other Deliverables
- Serviceable Design
- Manufacturable Product
- "Error-Free" Documentation

Figure 3.6-6 Development–test phase.

companies the proposed system must cross organizational jurisdictions and falls subject to a variety of communication barriers.

Today's highly competitive market is driving U.S. manufacturers and producers of high-tech products and systems to *plan* the integration of this life-cycle phase concurrently with the earlier chronological engineering design and development phases. "Quality by design" has rapidly become the theme and motif of such high-tech firms as 3M Corporation, AT&T, Hewlett-Packard, and Texas Instruments.

Systems analysts and hardware designers are encouraged to apply quality control methods to analyze system reliability early in the design process long before the project moves through manufacture. The belief is that the engineering–manufacturing–quality interface can be improved and the total product concept-to-delivery time reduced by attending to quality considerations in the design and development phases. If development is to hand off a product that is fit for use and free from errors, quality must be engineered into this product from the very beginning! Indeed, quality is conformance to specifications.

In other words, the development systems must be *manufacturable;* that is, the parts ordered should be usable in the manufacture of the product.

```
┌─────────────────┐
│   INTEGRATED    │
│  MANUFACTING    │
│  AND PRODUCTION │
└─────────────────┘
         │
         ▼
```

OUTPUTS
- Timely Procurement of All Required Resources—Inventory, Supplies, Funds, Labor
- Verification of All Production Specifications
- Volume Production Start
- Feedback to Developers and Manufacturing Engineers
- Component and Unit Testing
- System Integration and Testing
- Quality Assurance and Reliability Control or Compliance Testing
- Technical Manual and All Other Documentation Preparation and Production
- Product Installation–Deployment Support Plan Development

PERFORMANCE STANDARDS
- Manufacturable Product
- Established Cost Targets and Scheduled Milestones
- Established Performance Specifications
- Serviceable Design
- "Error-Free" Documentation

Figure 3.6-7 Integrated manufacturing and production phase.

(Often during the prototype development phase, parts are *carefully* selected to ensure that the prototype system will work. However, such a system is not manufacturable in a situation where, say, only 500 of the 1000 ordered parts do not work.) Errors identified here at the front end of the development cycle save significantly more than when found in the final product because errors incurred in succeeding life-cycle phases are much more costly to fix.

One of the best ways to design and develop a product for manufacture is to assign manufacturing personnel to the design and development teams. These individuals must do more than attend meetings and have coffee and danish. They must, instead, work with and become integral members of the product development team. Likewise, after transitioning to manufacturing, members from design and development should join the manufacturing team.

Integrated manufacturing and the simultaneous collaboration of manufacturing engineering in all subsequent production processes will:

- Shorten the traditional manufacturing life-cycle phase and subsequent product or system launch time.
- Reduce engineering changes after designs are introduced into production and frozen.
- Eliminate logjams caused by extensive production cross-checks.

Other quality improvement measures that can be taken by incorporating the following strategies into the production process are:

- Conformance to customer or user requirements.
- Training staff in the use of statistical quality control measurements.
- Just-in-time procurement.
- Process control engineering.
- Improved supplier performance.
- Timely new-product introductions.
- Reduction of the cost for not implementing quality assurance measurement.

3.6.4 Installation Phase

The primary operational emphasis shifts in this life-cycle phase from product planning to that of product control. Assuming that the product's design has been verified through preliminary unit and system testing in the factory, the Project Manager's duties now emphasize client interface because it is at this point that the product begins its commercial or sales' life cycle.

Typically, the following activities must be planned and carried out by various support organizations that interface with the firm's customers:

1. Negotiation of qualification sites for acceptance or market (beta) testing. (Beta testing is often done in the development phase as well.)
2. Preparation of all customer support services—documentation, training and installation manuals, repair services, etc.
3. Contract for delivery.

Planning and conducting final qualification site or marketplace tests of such complex, leading-edge technologies, for instance, as digital switching systems or office automation workstations is extremely important to guarantee customer satisfaction. Market (beta) tests measure the reaction of customers to an R&D product, service, or system. Such tests enable a firm to refine the design for subsequently planned releases and to revise the accompanying maintenance manuals and training documentation along with the final packaging and planned promotionals. Adequate market testing would have saved such high-tech firms as AT&T and IBM millions of dollars in planned

```
┌─────────────────────────┐
│                         │
│   INSTALLATION          │
│                         │
└─────────────────────────┘
                  │
                  ▼
```

OUTPUTS
- Documentation Delivered to Installers and Customers
- Trained Installation Support Personnel
- Trained Users and Customers
- Field Performance and Reliability Testing
- Product Release–Cutover at Pilot (Beta) Qualification Sites
- Acceptance Testing of Product by Users and Customers
- Customer Feedback to Product Planners and Developers

PERFORMANCE STANDARDS
- Established Performance Specifications
- Scheduled Field Test and Early Training Unit Shipping Milestones
- Serviceable Product Standard
- Accepted Product Qualification Evaluation Standards

Figure 3.6-8 Installation phase.

R&D activity following their launches of NET 1000 and the IBM PC Jr. Both of these products were withdrawn from production after repeated efforts by AT&T and IBM to "fix" what the customers told them was wrong. The output and standards of the installation phase are described further in Figure 3.6-8.

3.6.5 Divestment Phase

A project will live on forever if its timely completion is not planned, organized, or controlled. For this reason, some corporations often relieve a Project Manager of an assignment and appoint to that position a qualified individual who is especially skilled in closeout procedures.

The primary responsibilities of the Project Manager during project termination include:

- Developing a specific plan and schedule for redeployment of project personnel
- Ensuring that all required actions are taken to facilitate the customer's acceptance of all planned project deliverables

```
┌─────────────────────────────┐
│                             │
│     DIVESTMENT              │
│                             │
└─────────────────────────────┘ ▼
```

OUTPUTS
  • Monitored and Improved Product Quality
  • Monitored and Improved Product Market Position
  • Transfer of Resources to Product Maintenance or
    to Other New Projects
  • Evaluation of Total Project:
    – Performance Problem Assesment of Product and
      any Deviations from Original Specications,
      Planned Costs, and Schedules
    – Recommendations for New Development Efforts
    – Policy and Managerial Evaluation of Techniques
      and Procedures Employed on Project

PERFORMANCE STANDARDS
  • Established Closeout Schedule
  • Aquired Data Base of Product Defects, Reliability,
    Usability, Maintainability, Performance

Figure 3.6-9 Divestment phase.

• Assuring that the acceptance plan and schedule comply with contractual requirements
• Assisting Marketing in preparing a technology plan for the new product
  line that will not only introduce new products resulting from the
  continuation of this R&D effort but also enhance the market position of
  those already in existence by offsetting the effects of obsolescence or
  market saturation
• Conducting, along with the functional manager, "postmortem" evaluations or audits of the endeavor in an attempt to improve on future
  project management procedure and technique.

Figure 3.6-9 contains a description of the outputs and standards of the
divestment phase.

3.6.6 The Software Life Cycle

Although the life cycle of a software project is contained in the life-cycle
discussion presented above, it is useful to take a special look at software.

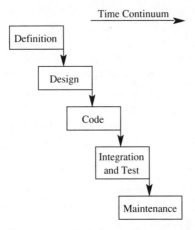

Figure 3.6-10 Software Life Cycle.

Software is required for all major projects and usually represents the *critical path* in theses projects (i.e., the duration of most projects is set by the software life-cycle).

The software life cycle consists of the phases shown in Fig. 3.6-10. The *Definition* phase is similar to that described above. We learn the requirements and plan the software project. The specifications are "firmed"; a preliminary design performed; and a work breakdown structure developed from which the software organization, budget, and schedules are obtained. In the *Design* phase, a detailed design is performed; the software system is divided into modules; interfaces are carefully specified, and test plans developed. In the *Coding* phase, each module is coded, tested, and debugged so that each software module meets the specifications for which it was designed. The Coding phase is completed only when each module operates without error (i.e., each module *conforms* to specifications). In the *Integration and Test* phase, the software modules are integrated together and then integrated with the system hardware. System testing and debugging follow to ensure that the entire system now meets specifications.

While systems and products limited to hardware go through a *production* phase, products comprised of both software and hardware have a *Maintenance* phase. The maintenance phase occurs after delivery. The customer gains familiarity with the system and "wants more". Often it's more software capability. "I want it to do this and that and . . .", is a frequent and pleasant refrain. However, modifying software, unless it has been modularized, is a major undertaking. Even when software is modularized, changing specifications means changing modules and testing software characteristics previously not planned. The result is uncovering new, undiscovered, bugs which must then be corrected.

We recommend the following to minimize "aggravation" in software maintenance:

- Understand *Specifications* fully, prior to design. Conversely, the design should include only those performance requirements which are necessary (i.e., that are contained in the specifications).
- *Design,* identify, and specify all modules and the connections between modules (interfaces). Be consistent in the design procedure.
- *Document* the code. Explain what you are doing. Two lines of documentation per line of code is superb. One is satisfactory. It is our experience that engineers writing software should code with only one-hand, and use the other to produce the accompanying documentation. Maintain records of all test results (an audit trail) which relates the software to the specifications laid out in the system requirements.
- Design for *Testability*. A code must be tested against specifications. Design so that the testing can be readily performed.
- Design for *Maintainability*. Remember whatever you code, even if it works well, the user will want more applications and, therefore, the code will be changed! Design so that change is readily achieved.

3.7 THE PROJECT SUMMARY PLAN

The Project Summary Plan is a summation of all the project's essential elements prepared in such a manner that management can learn at a glance what is to be done; why and how it is to be done; and when project deliverables are to be ready, by whom, and at what price.

It is necessary to also make these plan elements available to all project personnel from the very beginning to obtain not only their input but also their commitment to completing project objectives.

The components of a Summary Plan include:

- The Statement of Work (SOW).
- The Milestone Schedule.
- The Master Budget.
- The Work Breakdown Structure (WBS).
- The Project Network.
- A Task Responsibility Matrix.

The Statement of Work (SOW) is a narrative describing the work activities to be completed, the funding, and product specifications. All government proposals must be written and prepared in accordance with the specifications stated in this document. In fact, it is good practice for firms bidding on government contracts to reproduce the SOW in the body of their proposal response to indicate compliance with the customer's work requests.

The milestone schedule is a high-level schedule that should include such items as start and end dates of major work tasks or activities, along with tangible milestone events, report due dates, and the like.

The master budget is an estimate of funds required to complete each work activity over the life-cycle phases. The budget usually is presented as a schedule indicating the anticipated funds to be expended as a function of time.

The WBS breaks down the work activities into definite units or packages that can be assigned specific project or job numbers, expected cost, personnel, duration, and required equipment.

The project network is a schedule that shows the sequencing and interdependences of the work activities and indicates whether they will be accomplished serially or in parallel. The network indicates those activities on the critical path (i.e., those activities which, if delayed, will delay the project).

A task responsibility matrix or Linear Responsibility Chart (LRC), discussed in Chapter 2, illustrates who has been assigned the work and shows how the project organizations will interact when accomplishing the project objectives.

3.8 THE WORK BREAKDOWN STRUCTURE

One of the most important components of a project plan is the proper definition of the work tasks to be performed. A WBS, the primary planning and analysis tool used in engineering, answers two key questions:

- What is to be accomplished?
- What is the necessary hierarchical relationship of the work effort?

The WBS also:

- Provides a complete list of the work tasks consisting of the software, hardware, services, and data during the development and production of a product.
- Defines the responsibility, personnel, cost, duration, risk, and precedence of each work task.
- Provides an easy to follow numbering system to allow for a hierarchical tracking of progress.

This important planning and analysis tool has been successfully employed by federal agencies, such as the DOD, DOE, FAA, and NASA, as well as corporate R&D laboratories, software development companies, manufacturers, and civil engineering or facility management organizations. The aim is to better manage investments and projects of a variety of types and durations.

The WBS partitions the project into manageable elements of work for which costs, budgets, and schedules can more readily be established. When properly prepared and completed, the WBS satisfies the needs of management, the Project Manager, and the customer. The integration of the project's organizational structure with the WBS helps the Project Manager to assign responsibility for the different technical tasks to specific project personnel.

3.8.1. Preparing the WBS

Formation of the WBS family tree begins by subdividing, or *partitioning,* the project objective into successively smaller work blocks until the lowest level to be reported on or controlled is reached. This treelike structure breaks down the project work effort into manageable and independent units that will be assigned to the various specialists responsibile for their completion, thereby linking in a very logical manner company resources and work to be performed.

Figure 3.8-1 shows a level 2 WBS for a radio transceiver (transmitter–receiver). We see that the basic radio system listed in level 1 is partitioned with its main functional blocks in level 2. If done properly, there should be no work that must be performed that cannot be considered to be a part of

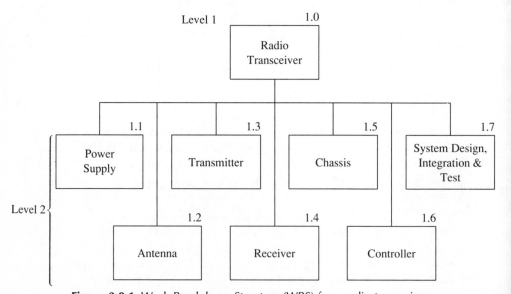

Figure 3.8-1 Work Breakdown Structure (WBS) for a radio transceiver.

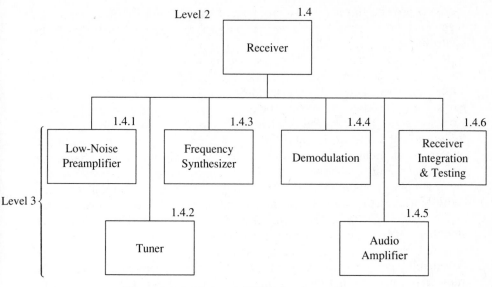

Figure 3.8-2 (*a*) Level-3 WBS for the receiver.

one of these level 2 blocks. Every cost expended or report written will list the task number assigned to assure proper credit.

The WBS is *not* a breakdown by discipline or functional organization, such as hardware, software, systems engineering, mechanical engineering, and so on, but as shown above, by functional blocks. Each of the level 2

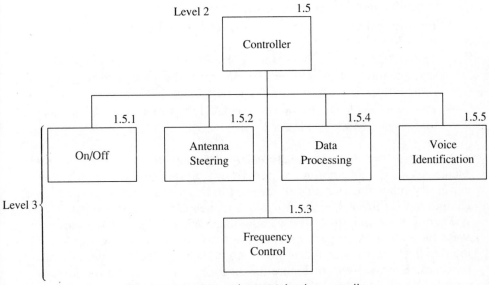

Figure 3.8-2 (*b*) Level-3 WBS for the controller.

blocks is now partitioned into a grouping of level 3 blocks. For example, the receiver block could be divided as shown in Figure 3.8-2(*a*) and the controller as shown in Figure 3.8-2(*b*).

Referring to Figure 3.8-2(*a*), we see that the receiver consists of a low-noise preamplifier, tuner, frequency synthesizer, demodulator, and an audio amplifier. These are the building blocks into which the receiver is partitioned. In addition, these blocks must be designed and fabricated to work together. This requires a further element of the WBS called *receiver–integration* and *testing*. Each of the level 3 elements can now be partitioned into more elements. The partitioning process ends when any element is well-defined and "buildable" by a small group of engineers. For example, the low-noise preamplifier, the frequency synthesizer, demodulator, and audio amplifier probably would not be partitioned further. However, one might further partition the *tuner* into a fourth level: RF amplifier, IF amplifier, mixer, and automatic gain control (AGC).

Referring to Figure 3.8-2(*b*), it is seen that the controller can be partitioned into five software elements. Each of these elements is eventually integrated into the hardware. For example, element 1.2 refers to the antenna. It should then be partitioned into a level 3 consisting of a physical antenna and antenna hardware/software integration and testing.

There is no "right" or "wrong" way to perform the WBS. Indeed, if, three people set out to develop a WBS for the same project, there would be three different WBSs. The important points to remember are:

1. Always prepare a WBS.
2. Do it by using a group of individuals having a diversity of expertises.
3. Be complete. Any element of the project against which funds are expended must be included in the WBS.
4. Be sure that the breakdown is a partition of building blocks and *not* a breakdown by organization or discipline.
5. The WBS is treelike. Therefore a block in level 4, for example, must separate into 2 or more level 5 blocks. All of the *work* performed in the level 4 block is performed in the level 5 blocks.
6. Two or more blocks, for instance, in levels 2, 3, or 4 can never connect to the same block in a lower level.

In practice the WBS is usually enlarged to include (at level 2) Project Management, Documentation, System Design, System Integration, Testing, and every other function that is *resolved* in the project and against which charges are incurred. Each level-2 block is then expanded as needed.

Figures 3.8-3 and 3.8-4 are two additional WBSs. Figure 3.8-3 shows a WBS for a software project while Figure 3.8-4 is a hardware project. Note the tree-like structure, and also that each branch divides into two or more branches. For example, in Figure 3.8-3, Task 1.3.3.2, *coding routines,* is completed when the level 5 tasks, *perform coding* (Task 1.3.3.2.1) and *unit*

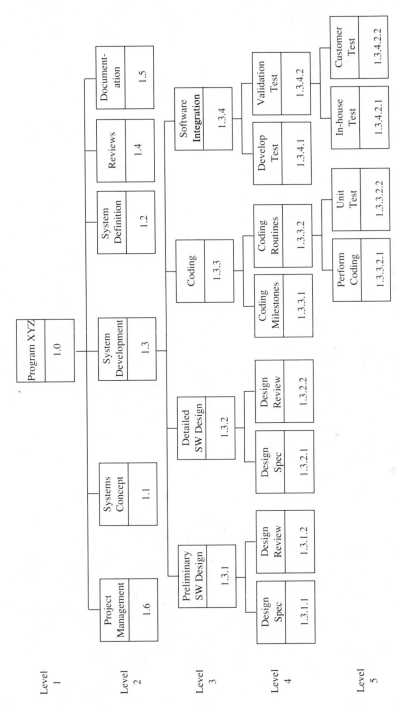

Figure 3.8-3 Work Breakdown Structure (WBS) for a software development project.

81

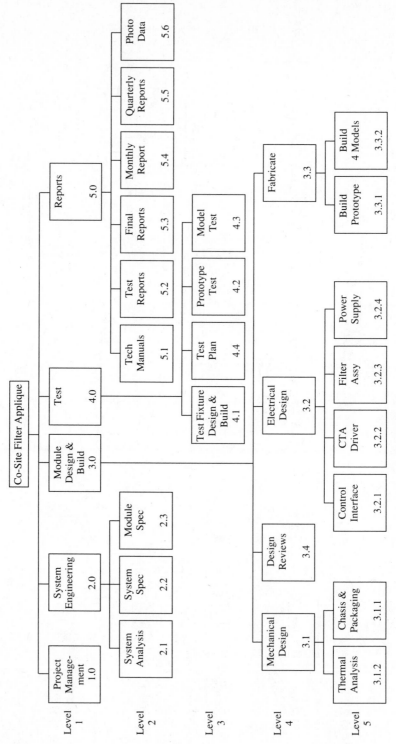

Figure 3.8-4 Work Breakdown Structure (WBS) for a hardware development project.

test (Task 1.3.3.2.2) are completed. In practice, each coding module would be delineated as tasks 1.3.3.X (X = 2, 3, 4, . . .).

3.8.2 The Work Package

After the WBS is constructed, the "finishing touch" is added. This is the "work package" or "dictionary." The work package is a description of what must be performed, by whom, and in what time duration.

The work package is always prepared for *each* bottom level element of the WBS. Thus, a level 2 element, if not partitioned (a highly unlikely possibility), would be given a work package; and a level 4 element, if not partitioned, would also be given a work package.

A typical work package is shown in Figure 3.8-5. Item 1 gives the project title, item 2 the WBS task (element) title and number, item 2a the number of times that this work package has been revised, item 3 the date that this work

Work Package

1. Project Title _____   3. Date _____
2. WBS Task (Element) Title / Number _____   2a. Revision _____
4. Project Manager _____   Telephone _____
5. Task Supervisor _____   Telephone _____

6. Specifications:

7a. Work description:

7b. Deliverables: 1.
                2.
                3.

8. Duration:

	t_o	t_p	t_m	Risk	Precedence
1.					
2.					
3.					

9. Personnel:

10. Special equipment / facilities / requirements:

11. Meeting schedule:

12. Reporting schedule:

Figure 3.8-5 A typical work package.

package was approved, item 4 the name and telephone number of the Project Manager, item 5 the name and telephone number of the Supervisor of the work to be performed for this WBS task, and item 6 lists all the specifications that this task must meet. These specifications might be enumerated in the list of product and project specifications, or they might have to be derived from these specifications by the Project Manager, the Task Supervisor of the next-higher-level task (which was partitioned and thereby includes the task being specified), or the Task Supervisor of this lowest level task.

Item 7 describes the work to be performed, such as design, develop, assemble, test, and integrate for a hardware task, or design, code, test, debug, document, and so on for a software task.

Item 7a lists the deliverables. For software, one might be required to deliver the design, the source code, the object code, all algorithms, test plan, test results, and any other documentation.

Item 8 lists the duration of time required to obtain each deliverable in Section 7a and the *risk* associated with each deliverable and the precedence relationship, that is, which tasks (regardless of where they are located in the WBS) must be complete before we can begin work on the deliverables. The task duration should be determined, when feasible, by the people who will be performing the task. One of the many reasons why projects are delayed (see Chapter 4) is that the Project Manager or a level 2 manager sets the duration of a level 4 or level 5 task. The probability is high that the level 2 manager last performed the level 4 task 10 years earlier and has forgotten the problems that ensued, remembering only the successes. The result is a very optimistic schedule. Three possible durations should be noted: an *optimistic* duration t_o, which is the time that the task would take if, indeed, "all went well"; the *pessimistic* duration t_p, which is the time that the task would take if "everything went poorly"; and the *mostlikely* duration t_m, which is the duration that you would expect on the basis of your experience. The authors recommend setting t_m midway between t_o and t_p initially and moving it toward t_o or t_p depending on your expectation of finishing earlier or later than this midpoint.

The risk will determine the spread $t_p - t_o$. A low-risk task is one in which $(t_p - t_o)/t_p < 15\%$, a medium-risk task is one in which $15\% < (t_p - t_o)/t_p < 50\%$, and a high-risk task is one in which $(t_p - t_o)/t_p > 50\%$. These numbers should be considered as guidelines, and actual numbers should depend on experience and the project.

Chapter 4 presents an in-depth discussion of the implications of item 8 on scheduling. The schedule presented with item 8 depended on a certain number and quality of personnel. Item 9, personnel, lists the number of personnel needed and their required qualifications. Using this list, the Project Manager is able to compute cost, since each member of the project team is paid by the project, depending on the number of hours worked on the project. Item 10 helps the Project Manager to ascertain cost and prevent delays by ordering all equipment up front, while items 11 and 12 indicate the meeting and report schedules.

The work package, when completed and agreed to by the Task Supervisor and the supervisor's manager, gives *both* the supervisor and the manager a complete description of the task to be performed. The work package should be initialed by both and kept available for easy reference.

3.8.3 Summary

In summary, each element of the WBS is partitioned until a work package can be meaningfully prepared. Reporting levels are numbered beginning at the total program or project level and illustrate a progression down to the lowest subdivision end item or work package. The integration of the project's organizational structure with the WBS occurs at this level. This point of integration of work effort and organizational accountability is generally assigned a cost accounting code and is used by management as a reference point to track and report expenditures.

To enable both Functional Managers and Project Managers to track resource allocations made to individual work assignments, the work packages contain information regarding cost, schedule, and technical performance requirements.

3.9 CONCLUSION

In Chapter 3 we have examined how projects and programs are planned using purposeful techniques that enable the project team to fulfill assigned responsibilities and to achieve overall corporate goals and objectives. Now that the project work has been properly planned, the next step is to execute that plan by building a schedule.

REFERENCES

1. Hax, A. C., and Majluf, N. S. *Strategic Management.* Englewood Cliffs, NJ: Prentice Hall, 1984, pp. 37–41.

BIBLIOGRAPHY

Abetti, P. A. "Milestones For Managing Technological Innovation." *Planning Review,* (Vol. 13, March 1985, pp. 18–22, 45–46.

Archibald, R. D. *Managing High-Technology Programs and Projects.* New York: Wiley, 1976.

"Blind Spot in Strategic Planning?" *Management Review,* October 1984, pp. 26–28, 49–52.

Cleland, D. I., and King, W. R. *Systems Analysis and Project Management* (3rd ed.). New York: McGraw-Hill, 1983.

Cudworth, B. F. "3M's Commitment to Quality as a Way of Life," *Industrial Engineering,* Vol. 7, No. 1, July 1985, pp. 16–17.

Harrison, F. L. *Advanced Project Management,* Huntingdon, U.K.: Cower Publishing Company Limited, 1981.

Hayes, R. H. "Strategic Planning—Forward in Reverse?" *Harvard Business Review,* Vol. 63, No. 6, November–December 1985, pp. 111–119.

Hayes, R. H. "Why Strategic Planning Goes Awry," *The New York Times,* April 20, 1986.

Kerzner, H. and Thamhain, H. *Project Management for Small and Medium Size Businesses.* New York: Van Nostrand Reinhold, 1984.

Morris, P. "Managing Project Interfaces—Key Points for Project Success," in Cleland and King (Eds.), *Project Management Handbook.* New York: Van Nostrand, 1983.

Putnam, A. O. "A Redesign for Engineering," *Harvard Business Review,* Vol. 63, No. 1, May–June 1985, pp. 139–144.

Raelin, J. A., and Balachandra, R. "R&D Project Termination in High-Tech Industries," *IEEE Transactions on Engineering Management,* Vol. EM-32, No. 1, February 1985, pp. 16–23.

Richardson, P. R., "Managing Research and Development for Results," *Journal of Product Innovation Management,* Vol. 2, No. 2, February 1985, pp. 75–87.

Case Study

The IRSS Development Project

Introduction

What had been perceived some nine months ago as an excellent product for launching Lutech Systems into the medical communications market was clearly in jeopardy.

Ron Lewis, Vice President of Lutech's newly created Medical Diagnostic Planning Division, studied the results of the recently commissioned project audit. Again and again, he asked himself, "How in the world has this team of qualified designers performed so poorly! What had gone awry to cause a 9-month slippage? Could I now trust their projections to resolve the identified problems?"

As Ron pondered the report, he realized that whatever plan he implemented to correct the situation, he'd need the support of all his technical managers, Marketing, and the respective general managers from Manufacturing. To keep the project viable, he'd need to improve the interfaces between these organizations as the contributions of each one of them was paramount to the success of Lutech's new endeavor.

What Went Wrong?

What had been conceived of in the laboratory as a single application program, for a larger software system, eventually became more complex in direct assessment of the market potential. Over the past decade, medical electronics had advanced diagnostic medicine in the United States and abroad. Numbering among the modern marvels is sophisticated equipment that displays magnetic images of the human body's changing chemistry, as well as ultrasonic scanners and infrared sensors that use sound waves and thermography to provide pictures of vital human organs. With these advances in medical diagnostic imaging, there was a push for an innovative system that would make possible the storage and retrieval of patient data, general medical information, and image archiving. Over the next 10 years as the major public and private hospitals and medical institutions purchased several of these anatomical machines, their needs for a comprehensive interface system would grow.

Lutech's executives concurred with the market assessment and funded the communication program's development. In fact, they went so far as to carve out a Medical Diagnostic Planning Division at Lutech and staffed the

---

This study was written for class discussion purposes only. It is not intended to reflect management practices or products of any particular company or organization.

organization by recruiting talent and expertise from the country's top graduate engineering schools. All this was done in anticipation of launching several prosperous endeavors within the next 2–3 years.

But it soon became apparent to the Information Retrieval and Storage System (IRSS) project auditors that the new developers were inexperienced program planners. The communication program was described to them as having the complexity of an operating system. The program would direct the operations of the many hardware devices and operate on any number of host computers. However, the preliminary project plan included only a broad definition of the program's capabilities; it contained no goals or requirements pertaining to the project itself. There were no controlling specifications for testing or debugging the software protocols and the user interfaces, nor for determining recovery time in the event of system failure. Also lacking were detailed plans for writing concurrently the program and installation documentation, although the preliminary plan included two technical writers for the final stages of IRSS's development. And nowhere in the plan was there reference to beta test site installations where the accompanying hardware system's design would be tested.

The project had been operating in a matrix organization, where Functional Managers had retained responsibility for their own budget and project resources. To date, there was no unified effort to staff IRSS as a single project or to dedicate full-time resources to it.

The Project Organization

Although the number of specialists completing project work spanned over three departments in the new Medical Diagnostic Planning Division, project specialists needed to interface with five or six other groups dispersed throughout the company. The manufacturing site, furthermore, was remote to the location of the three development and technical documentation departments. The separate work groups proceeded independently, with no central management directing coordination of the interfaces or assignment hand-offs.

Hence there was a tendency to overcommit and a great possibility for error. To make matters worse, the development organizations would fail to provide Marketing with what the customers requested. The product's lack of systems engineering and inavailability of a detailed development schedule contributed to the IRSS designers' failure to make solid commitments. Just as important, the absence of a Project Manager to work within the matrix resulted in unclear management roles, turf tissues, and conflict regarding project deliverables.

In a very real sense, the IRSS work effort was not a "project"; nor would it ever become a true commercial venture until the lack of direction among the multidisciplined work efforts was rectified.

QUESTIONS

1. What would you recommend to the new Vice President as a course of action for clarifying project objectives?

2. Name at least four reasons why these top-notch designers might *not* succeed in developing the IRSS.

3. What general indications exist that Lutech System's present organizational structure may not be adequate for a project of this size and scope?

4. How have communication barriers among the interfacing project groups contributed to the difficulties cited in the audit report?

The Vice President's Plan

Ron considered carefully whether to salvage or abandon the project and, after much deliberation, determined that IRSS should be restructured in ways that would ensure commitment. He knew that establishing new policies regarding the project organization was a key. Armed with the auditor's report, he organized a central management project team, and then named a Project Manager to oversee their activities as well as those of the designers assigned to develop the medical information system and each of the several technical support groups. He charged the management team with the responsibility of planning the project and authorized the direct transfer of required funds from the functional departments involved in the work to the project organization. Ron also requested that the Project Manager keep him informed on a regular basis of progress, problems, and the team's examination of all alternative courses of action to correct problems. Ron pledged to reward progress and encouraged candid reporting in an attempt to ensure that the management team would attend to the quality of problem analysis and decision-making.

Directions To The Project Team

The morning kickoff meeting of the new IRSS project management team with the Medical Diagnostic Planning Division Vice President was the first of many. Ron Lewis insisted that his new managers be informed from the project outset of the company's general thrust in the biomedical high-technology market, as well as understand the importance of coordinating tight budgets and development schedules. Because of the complexity of the technical aspects of the project, they also would have to closely monitor the phased execution and testing of the product's design.

The new IRSS management team was made up of representatives from Lutech's Marketing and Manufacturing Divisions and a core of technical first-line managers who represented each of the project's technical disciplines. The Project Manager, Bob Anderson, was a seasoned systems engineer with an expertise in digital image processing. Ron Lewis hand-selected Bob for the position after carefully reviewing several other qualified applicants. Bob had been with Lutech Systems for 7 years and served as a Lead Systems Analyst on a number of development project under contract with the U.S. Department of Navy. Hence, Bob knew how to run a project. Given full reign to organize, plan, and control one, it was anticipated that he would succeed for the Medical Diagnostic Planning Division. In fact, the very first thing Bob did was to organize the project structure and create the framework of an overall technical and management plan—the details of which the project team would determine together.

A Project Structure Emerges

No one had promised Bob that his job as the IRSS Project Manager would be easy. He and the management team met at length for nearly 3 weeks to determine project objectives. They set out to size and estimate the extent of the anticipated work by breaking it up into manageable packages that were to be assigned to individual technical specailists. Anticipating a strong market demand for the product early on, they scheduled a phased release of the image-retrieval storage system obtaining in advance several medical research institutions as qualification sites to test the system. Once the initial objectives were decided on and a work breakdown structure prepared, it was necessary to obtain consensus over assigning the work outlined in the WBS and scheduling the technical milestones. Administration and tracking of the work package became the responsibility of Shara Jefferson. Shara joined Bob as Deputy Project Manager and organized a Project Office. Hence, the finally agreed-on schedules were displayed, tracked, and updated.

Shara was required to report on the activities of the design organizations. To monitor progress, she would circulate among the design groups asking questions, coordinating interfaces, and determining progress on the work packages. Through her hard work she gradually gained the trust and respect of the first-line managers. She also became a sounding board for several of the engineers. She would listen to their problems and advise them on all administrative project details. Early on Shara appraised Bob that the initial design specifications were inadequate. In fact, two approaches were initiated by different work groups. And the developers of both approaches were proceding unaware of their duplicated efforts.

Having completed their first report, Bob and Shara sent it to Ron Lewis for his review. Later that week, Ron called Bob and Shara to his office to discuss their findings:

"A lot of good has been done here. But we can't have these duplicate efforts holding up progress. Even though some of the design groups may resent this, a much more formal structure is going to have to be instituted to manage design changes and control the work effort. Executing a system this complex depends entirely on how well we coordinate its design every step of the way—from its robustness in the conception phase to user acceptance when deployed in the marketplace.

I'd say our job is clear—institute a mechanism whereby all the team members are informed of the other's activities and their difficulties or problems, whether technical or administrative.

Have them meet face to face and hass out their differences, if need be, on a weekly basis. I want each of these groups to learn from the other's mistakes and resolve their differences as a team solving the same problems. As Project Managers, you should run the meetings and keep lists of the unresolved issues. Demand weekly status reports from anyone having any responsibilities for work on this project. I want to be kept informed of any impeding delays before they happen!''

QUESTIONS

5. What recommendations would you make to the project management team for overcoming the many technical barriers facing them?

6. Why is it important to synchronize the different design and development activities?

7. What mechanism would you enact to provide guidelines for documenting designs and driving their execution?

Documenting Change

The team was undecided on the appropriate hardware needed to implement the software program and act as the interface with the many proposed host computers and diagnostic devices. By the time Bob and Shara staged the project's first technical design review, the hardware team had, in fact, operationalized an approach that used a high-speed microprocessor to perform the interface function. Although a most feasible solution to the design problem, the software designers had not yet completed the system's architecture.

Of additional concern, to the hardware team was the design of the cabinetry that would house the IRSS unit. Recommendations also would need to be made regarding the placement of the unit in the hospitals and medical institutions. The system's circuitry would need to be shielded from the powerful magnetic fields of any magnetic-resonance imagers physically linked with the unit. Other technical issues discussed at the meeting were the problems associated with the operation of the central data bases and auxillary storage units. In the event of trouble or failure of any one of the central computers, physicians and other medical personnel relying on the data would be inconvenienced. To minimize this risk, the feasibility of configuring the system for distributed data processing was considered.

Several of the software engineers argued for including a relational data base manager at the core of the program unit to store patient data and retrieve medical records. The pros and cons of this type of architecture were debated at length, with arguments pointing to the need to document all the proposed changes to the original design specifications in a project handbook—a record Shara and Bob would maintain "on-line" in the Project Office.

At first the project team resisted the idea of documenting every aspect of their work, but with the persistent coaching by Bob Anderson, they agreed to implement the more formal structure on a trial basis. He assumed the responsibility of writing and maintaining the high-level project information—project requirements, project description, and a detailed accounting of all the agreed-on hardware and software interfaces. The remaining sections were to parallel the hierarchical structure of the work packages so that the different design teams were responsible for maintaining and updating this information. The actual technical specialists were free to write individual memoranda as needed regarding the specifics of their assignments. These notes might include reasons for rejecting a design, design assumptions, and special problems imposed by memory layout or other features of the design.

Shara would continue to arrange and report on the results of the many design reviews and update and track the associated documentation through the product's development and final testing. Also, her presence at these review meetings would help to assure that the integrity of the high-level design was not compromised in the effort to carry out these requirements. Finally, a formal sign-off of approved documentation for each of the different project phases was instituted as a means of synchronizing the hardware and software development and testing processes as well as planning out the manufacturing intervals for the product.

What the IRSS management and development teams proved was that they could work together using a flexible yet formal methodology that accommodated their different work styles. The project handbook was structured to complement the project's work breakdown. It thereby served the team members directly, as the project information was accessible to all involved and was complete and accurate.

By the time the development team was ready to cut over their designs for manufacture, Shara had implemented a system whereby the project documentation was accessible "on-line" at the factory, along with the updated and baseline schedules of the IRSS development milestones.

Everyone associated with the project was extremely anxious to perform well. The key had been to get all parties working on the same problems and to establish a climate that stressed open and honest communication in the reconciliation of differences. The managerial actions having the greatest impact on the project's design and schedule took this into account.

QUESTIONS

8. Explain the impact of documenting changes in the program's design specifications on the overall success or failure of this project.

9. How did the management team go about its task of facilitating the resolution of undetermined design issues?

10. Why was it essential that the many engineering interfaces be formally integrated into one project plan that was agreed on and signed by all responsible parties?

11. Discuss the critical role played by the Vice President in expediting the project schedule and mediating problems that arose from the conflicting design philosophies.

EXERCISES

1. Explain why each of the items listed in Sec. 3.4 could cause a plan to fail.

2. What do *you* mean by a *plan failing?* What would you do if you see that a plan is in the process of failing?

3. Many managers and engineers do *not* approve of *contingency planning.*
    a. List and explain possible drawbacks of Contingency Planning.
    b. We prefer the use of Probabilistic PERT (Sec. 4.5) to Contingency Planning! Explain why.

4. Explain why we call *time, specifications* and *resources* the "triple constraint."

5. You are the Project Manager of a project. You believe the project will be late since there is 9 months left to be *on time,* but your schedule indicates 13 months are needed to complete the project. Your customer wants the product *on time.*

    List the possible options. Discuss their consequences.

6. You are the Project Manager. One of your engineering groups is trying to perfect its subsystem beyond what the specification requires.

Explain how would you handle the situation if the subsystem is:

    a. on the critical path;

    b. is not on the critical path.

7. Engineers and managers often state that they cannot estimate the cost of a R&D project since they have not yet done the work.

    a. List the advantages and disadvantages of such a statement.

    b. What are some of the techniques that should be used to estimate cost.

8. On many projects only the PM knows the true schedule and has a copy of the project plan. List advantages and disadvantages of such a procedure.

9. A Work Breakdown Structure can be prepared for "home-projects" as well as corporate projects. Consider planning a two-week trip to Europe. Design a WBS having no more than 5 levels.

10. You are in charge of planning a New Year's Eve party for 50 guests. Prepare a 5-level WBS.

CHAPTER 4

Scheduling Project Activities

4.1 INTRODUCTION

In this chapter we describe the principal scheduling techniques that we have found to be effective in planning and later in tracking and controlling the project's performance.

In Chapter 3 we represented the project by a Work Breakdown Structure (WBS). The WBS was divided into a sufficient number of levels to allow each task to be performed to be specified in terms of *resources,* such as personnel, equipment, facilities, and *time.* In addition, the *precedence relationships* between the tasks were also specified, as were *milestones,* or deliverables such as tests, reports, and receipt of equipment needed for the project to proceed.

With this information available, one can schedule the project. But first we ask one additional question—*whose* schedule shall we prepare? For example, should a schedule be prepared for the *programmer* coding a particular function, for the *supervisor* of several programmers, or for the *Project Manager?* While each requires a schedule, does each require the same schedule?

4.2 DIFFERENT SCHEDULES FOR DIFFERENT FOLKS

Different levels of management require a different level of schedule detail. Indeed, it is not unusual to hear a group leader describing the weekly progress report in the following way:

> The engineering supervisors each collect a one page progress report from their engineers and computer scientists. They then summarize these reports on one page which is then given to me. I check to see how each of their jobs are progressing by comparing their progress to the

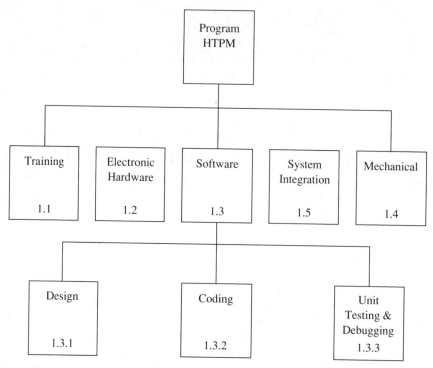

Figure 4.2-1 Partial WBS.

planned schedule. Then I summarize these results into a one-page report and give it to my manager. My manager receives seven reports each week, summarizes them, and submits a one-page report up one more managerial level until it reaches the Project Manager.

Each summarization of a group of reports and the subsequent passing of the report upward represents passage of the WBS from level i to level i-1.

Consider the partial WBS shown in Figure 4.2-1. If Bob Clark is in charge of the software aspect of the HTPM project, then he is extremely interested in viewing the details of the schedules derived by Sue Kaplan (1.3.1), Roger Pook (1.3.2), and Sandra Li (1.3.3). However, Gil Nolan, the Program Manager, is not interested in these details. Gil wants summary reports from Bob Clark (1.3) and the other managers of tasks 1.1–1.5 so that he can ascertain the schedule and the project performance.

The schedules that we discuss below can be obtained for any of the WBS levels. We recommend that you fully understand the reason for the schedule and for whom it is intended before preparing one.

4.3 WHY SCHEDULE?

Schedules are required as part of any project plan. The most important schedules are *performance* schedules, *personnel* schedules, and *cost* schedules. *Performance schedules* inform the project participants when each task is expected to begin and end. *Personnel schedules* inform each participant when each project member is expected to start working on each task, how many hours per day are expected to be allocated to each task, and when that project member is no longer needed on a given task. Personnel schedules allow us to determine whether anyone is overloaded and to see who is underloaded. By properly placing personnel (this is called *leveling* and is discussed further in Section 4.10.1), we can often shorten the project performance schedule and reduce cost. *Cost schedules* inform the Project Managers how much money has been allocated and spent for each task as a function of time. In this way adequate money can be requested prior to starting the project. All too often we find that projects end up in difficulty because adequate funding was not requested and allocated *up-front*.

4.3.1 Planning a Schedule to Meet Required Constraints

In the planning phase we rarely find that our initial schedule estimates are adequate. The project may have a defined completion date, of, for example, March 13, 1989, yet the performance schedule estimate shows a completion date of December 16, 1988. Or a fixed cost of $2,000,000 might be available to complete the project, but our *cost* schedule estimate indicates that the project will cost $2,500,000.

In general, there is a *triple constraint: time, cost, and technical specifications*. All three factors seldom can be specified simultaneously. If one is decreased, the other two may have to be increased. Thus, if our project is *time limited* and we see from our three initial schedules that we cannot make the time limit, we must perform "what-if" analyses, that is simulations of other design strategies, other precedence relationships between tasks, and other allocations of resources to obtain a performance schedule consonant with the time limit. If the project is *cost limited*, we might attempt to reallocate resources and the task durations so as to keep the cost within the prescribed limit while usually minimizing time.

Unfortunately, we often find that when good scheduling techniques are not employed, it is only near the project's planned completion date that the Project Manager realizes that the project cannot be completed on time and that when completed, the project will be above cost. To meet the required deadline, the project specifications must usually then be weakened, often by removing *features*.

4.3.2 Scheduling for Control

Schedules can also be used to *control* the project's progress. The *performance* schedule, prepared in the planning (proposal writing) phase, should be compared monthly or biweekly to the actual project performance. Slippage should be noted and corrected when appropriate (see Chapter 5). The *personnel* schedule will let the Project Manager know who can be shifted from one task to another in order to improve performance. The *cost* schedule provides a means of notifying all managers how much the project actually costs, as compared to how much money is budgeted or set aside. In this way, it is possible to predict if extra funding will be required to complete the project on time, or how long the project might take to complete for a given level of funding. This procedure is called the *Earned Value Analysis* (EVA) and is discussed in detail in Chapter 5.

4.4 THE PRECEDENCE (PERT/CPM) DIAGRAM

Consider a simple project ABC which consists of activities shown in Table 4.4-1.

The first question on every manager's mind is "How long will this project last?" The next question each manager usually asks is "When will my people be needed?" These two questions, and more, are answered through the use of a *precedence* or *PERT (CPM)* diagram.

Precedence diagramming is a technique in which each activity is represented by a rectangle. Each rectangle is then connected to the rectangle or retangles representing activities that *succeed* it in time. The first rectangle is

Table 4.4-1 Work Breakdown for Project ABC

Activity Designator	Activity Description	Proposed Activity Duration (months)	Immediate predecessor
A	Design, code, and test software	14	Start of project
B	Design, develop, and test hardware	3	Start of project
C	Integrate hardware and software	3	A, B
D	Design and fabricate chassis	7	B
E	Integrate and test system	4	C, D
F	Deliver system and perform customer test	10	E

usually considered *start* and the last rectangle, *finish*. Figure 4.4-1 represents the precedence diagram for the project described in Table 4.4-1.

Note, that to obtain the diagram in Figure 4.4-1, we drew a rectangle to represent each activity *A* through *F*, as well as rectangles to represent *start* and *finish*. These rectangular boxes are then connected by directional lines, shown as arrows, to indicate the precedence relationships.

The utility of the precedence diagram lies in its ability to simplify the calculation of the project's *duration*. Because some tasks are performed in parallel, the duration is less than the sum of the times it takes to complete each activity *A* through *F*. Indeed, in this problem, the project's duration is the time it takes to finish tasks *A, C, E,* and *F; B, C, E,* and *F;* or *B, D, E,* and *F*. The durations of these tasks are:

ACEF: 31 months
BCEF: 20 months
BDEF: 24 months

Because the project cannot be completed until *each activity is completed,* the project's duration is determined by activities *A, C, E,* and *F* and is 31 months. Note that in 31 months all activities will be completed. However, as a result of the precedence relationships, we cannot complete the project in less than 31 months.

4.4.1 Critical Path

The path *ACEF* (see Figure 4.4-1), which determines the project's duration, is called the *critical path*. If any activity on this path (i.e., *A, C, E,* or *F*) is delayed, the entire project will be delayed.

If the duration of the project, as determined by the precedence diagram, is too long (e.g., *Marketing* has determined that the project must be completed in 28 months or the *product entry window* will be missed), then it is clear from the precedence diagram that one or more of the activities on the critical path must be shortened. Several methods for shortening the project are:

1. Examine whether an activity on the critical path can be started sooner. For example, refering to Figure 4.4-1, let us see if activity *C* can be started earlier. We could shorten the project by integrating the hardware and software while performing the software test and debug phase of activity *A*. It is often possible to subdivide an activity such as *A* into two or more activities, for example, into A_1, A_2, and A_3, so that *C* can be started after A_2. However, *C must start after B* and the software coding are both completed. If we make these changes in the precedence diagram, Figure 4.4-1 can be redrawn as illustrated in Figure 4.4-2. Referring to Figure 4.4-2, we see that the critical path is

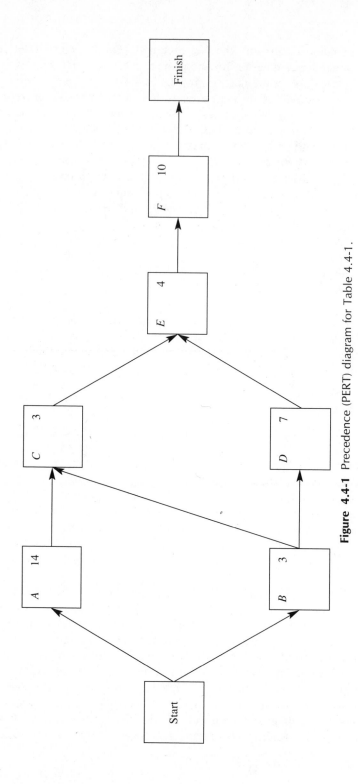

Figure 4.4-1 Precedence (PERT) diagram for Table 4.4-1.

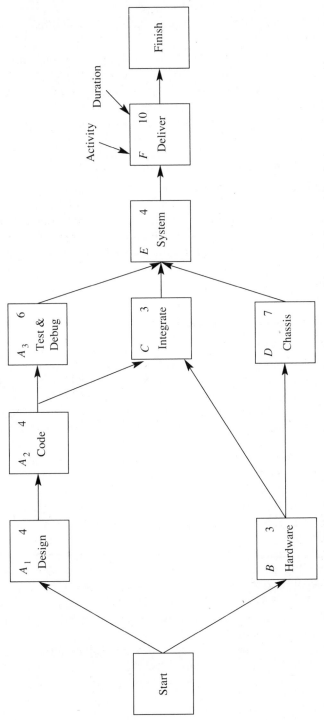

Figure 4.4-2 Precedence diagram for project *ABC* after splitting activity *A*.

now $A_1 A_2 A_3 E F$ and the project duration is 28 months. We have met the required product window.

2. Determine whether a different technology could be employed that could possibly shorten the time of a given activity. For example, in project *ABC* perhaps some of the software could be replaced by hardware, for instance, a *programmable array logic* (PAL). Because hardware development is not on the critical path, increasing *B* by a modest amount will not affect the project duration, while decreasing *A* will actually shorten the project. Of course, one must check the precedence diagram to see if the critical path has been changed.

3. Determine whether increasing the number of programmers in *A* will reduce its duration. This technique is discussed in detail in Section 4.8 but at this point it is useful to note that this procedure is of limited value. For example, doubling the number of personnel assigned to an activity rarely halves its duration. Indeed, addition of staff may at times *increase* the duration of an activity since it increases the amount of communications required.

4.4.2 Calcuation of the Critical Path

In the example shown in Figure 4.4-1, the critical path can be found by determining each and every possible path, and then determining which path is of longest duration. That path is, of course, *critical*. However, in any real project there may be 100 to 1000 or more such paths and the determination of the critical path would be an extremely difficult and time-consuming task. Further, because we calculate the project duration from the critical path in order to obtain an indication of what tasks should be shortened, or what precedence relations should be altered, we generally recompute the critical path several times before finalizing the precedence relations, resources per task, conceptual design, and so on.

To recompute the critical path several times in the tedious manner described above is not impossible but is so unpleasant a chore that an alternate procedure has been found. This procedure employs the *dynamic programming algorithm* and is readily explained by using Figure 4.4-2.

Referring to Figure 4.4-2, we see that prior to starting task *C* we must first complete tasks A_1 and A_2 and then complete task *B*. As tasks A_1 and A_2 take 8 days to complete while task *B* takes only 3 days to complete, task *C* can start only after 8 days have elapsed. Tasks A_1 and A_2 are therefore *critical* so far as task *C* is concerned. We can then, to all intents and purposes, ignore, or "cut," the path from task *B* to task *C* when calculating the critical path.

Next, examine the tasks in Figure 4.4-2 that must be completed prior to starting task *E*. They are tasks $A_1 A_2 A_3$, tasks $A_1 A_2 C$, and tasks *BD*. Note that we have omitted *BC* as we have already shown that *BC* is *not* a critical path. Since $A_1 A_2 A_3$ is the critical (longest) path to *E*, we can *cut* path *CE* and *DE* in determining the critical path.

There is now only one path left: $A_1 A_2 A_3 E F$. This is the critical path. Of most importance is the fact that the critical path was found by computing the longest path through the network. All computer-based project management systems operate in this fashion.

Example 4.4-1

A new system intended to revolutionize the office work area was designed and built using the following size of tasks (work breakdown):

	Task	Duration (months)	Precedence
100	Hardware design	4	Start of project
110	Breadboard construction	8	100
120	Test breadboard and correct	8	110
200	Design prototype	2	120
210	Develop prototype	6	200
220	Test prototype and correct	8	210
300	Design software	6	Start of project
310	Code all software	4	300
320	Test and debug software	6	310
400	Design mechanical system	3	200
410	Fabricate mechanical system	3	400
500	Integrate electronics and mechanical system	2	220, 410
510	Test and correct	3	500
600	Integrate software and hardware	1	510, 320
610	Test and correct	2	600
700	Prepare initial hardware documentation	1	120
710	Upgrade documentation	1	700, 320
720	Prepare final documentation	1	610, 710
800	Forward to customer for beta testing	1	720
	Finish	—	800

Problem

1: Prepare a precedence diagram.
2: Calculate the critical path.

Solution
1. The precedence diagram is shown in Figure 4.4-3.
2. To calculate the critical path, we first compute the elapsed time to

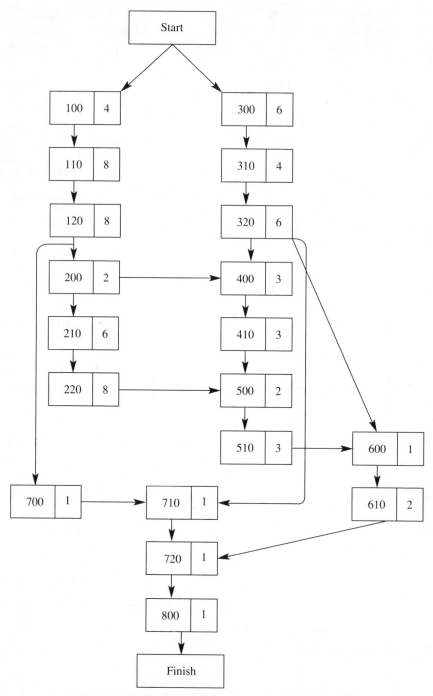

Figure 4.4-3 Precedence diagram for Example 4.4-1.

complete tasks 100–220. The time is $T_1 = 36$ months. We next compute the time to complete the tasks 100–410. This time interval is $T_2 = 28$ months. Because T_1 is greater than T_2, tasks 400 and 410 *cannot* lie along the critical path. This result can also be seen by comparing the durations of tasks 210 and 220 with tasks 400 and 410. Since tasks 210 and 220 take 14 weeks compared to tasks 400 and 410, which take 6 weeks, tasks 400 and 410 are seen to *not* be critical. Indeed, it is readily shown that task 400 can be started as early as week 22 or as late as week 30 and task 410 as early as week 25 or as late as week 33. This ability to start tasks 400 and 410 at any time over an 8-month period is called "slack" or "float."

Next, we compute the duration of tasks 100–510 and compare this result to the time interval needed to complete tasks 300, 310, and 320. The result is:

Tasks	Duration	Possible Critical Path
100–220 to 500–510	36 + 2 + 3 = 41	Yes
300–320	6 + 4 + 6 = 16	No

Hence Tasks 100 to 220, 500, and 510 still appear to be critical. The next node having two or more inputs is task 710. Comparing the durations of the tasks involved yields:

Tasks	Duration	Possible Critical Path
100–120 and 700	21	Yes
300–320	16	No

Thus, since both paths leaving task 320 indicate noncritical paths, we are now sure that tasks 300, 310 and 320 are noncritical.

Entering node 720, we again have two possible critical paths:

Tasks	Duration	Possible Critical Path
100–220 and 500–510 and 600–610	36+5+3=44	Yes
100–120 and 700 and 710	20+1+1=22	No

The critical path is therefore comprised of tasks 100, 110, 120, 200, 210, 220, 500, 510, 600, 610, 720, and 800. The total project duration is 46 months.

If, in Example 4.4-1, a duration of 44 months is greater than desirable, one would investigate possible alternatives to the precedence relationships postulated. For example, we would investigate doing more of the tasks on the critical path in parallel. This would reduce the duration of the critical path. Further, we would investigate the possibility of adding personnel to tasks on the critical path so as to reduce the duration of those tasks. Then, we would attempt to determine whether any of the tasks *on the critical path* could be shortened by employing a different design or by using new technological aids, such as a faster computer, additional PCs, and an automatic circuit tester. Finally, one should give thought to modifying the project's specifications or features, as it is quite possible that a minor change in the specifications could significantly shorten the project's duration.

Often, these procedures are sufficient to shorten a project as required to meet marketing goals or contractual needs. If, however, no modifications can adequately shorten the project, this information is of great benefit *before* we begin an expensive program that must be completed on time or be doomed to failure.

There is very strong feeling among industrial and government leaders that timeliness in a product's delivery is most important. The Department of Defense has found that many of its military projects are late, often because of changes in the specifications that increase the scope of the work. An interesting, fictional, story is told of a *navigation* system that was to be built by a major engineering firm for the DOD. At the first design review meeting the contractor and customer agreed that the system, with a "minor modification" could also be used to send data. One year later, seeing that this modification appeared to work well, the contractor and the customer agreed to modify the system again, in a "minor" way, so that *voice* could also be transmitted. When the system was finally delivered, one year late and $500,000 over cost, it had *no* navigation system. When asked what had happened to the navigation system, the contractor or customer explained that the system became so complex that it had to be removed.

4.4.3 Activity on Arrow (AOA)

An alternative approach to drawing the precedence diagram is the technique called *activity-on-arrow* (AOA). In this section we describe the AOA technique, which is used considerably. However, we recommend against its use when preparing a precedence diagram manually because it is a more complicated tool than the activity-on-node (AON) approach, discussed earlier.

To illustrate AOA, consider the problem shown in Table 4.4-1. When using AOA, each activity is shown by an *arrow* having a *node* at each end. The arrow represents the activity, *not the node*. The first step in drawing the AOA is shown in Figure 4.4-4(*a*). Note that an arrow is drawn to represent each task. Next, in Figure 4.4-4(*b*) the arrows are connected by *dashed arrows,* following the precedence relationship given in Table 4.4-1. These dashed arrows are called "dummy" tasks and are of zero duration. The final

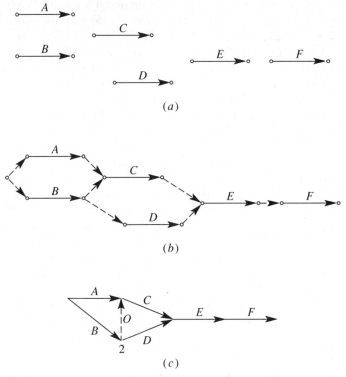

Figure 4.4-4 (a) First step in drawing the AOA; (b) introducing the dummy task; (c) the AOA precedence diagram.

AOA diagram, shown in Figure 4.4-4(c), contains a single dummy task. All other dummy tasks have been removed without affecting the required precedence relationships. Note that if the remaining dashed arrow is removed, by either erasing the arrow or by combining nodes 1 and 2, the precedence relationship is altered. It is this third step that complicates the use of the AOA. If we stop at Figure 4.4-4(b), the diagram is overly complicated, and if we go to Figure 4.4-4(c), the chance of error increases.

Thus, we recommend using AON for manual preparation of a precedence diagram.

4.5 PROBABILISTIC PERT: INCLUDING TASK DURATION, RISK, AND UNCERTAINTY IN A PRECEDENCE DIAGRAM

Probabilistic PERT (P-PERT) is a technique for including the risk and uncertainty inherent in the completion of every activity in a project. These activities are shown in the network precedence diagram (PERT diagram) that characterizes the project. If we are involved with the manufacture of "widgets," which have been manufactured for the past 10 years and have sold 3 million units, then we can accurately tell a customer how long it will take to fill an order for 10,000 additional widgets. If, however, we are involved in a project involving *research* and *development,* or the *manufacture* of a new product, then the time schedule for each activity required to complete the project or manufacture the product, is *uncertain.*

This *uncertainty,* or risk, in the completion of an activity can be expressed mathematically if we consider the *duration of each activity to be a random variable.* Hence, the duration of the critical path in the precedence diagram can also be viewed as a random variable because it represents the sum of random variables. Further, it is well known that the sum of random variables has a probability distribution, which approaches the Gaussian probability distribution. This theorem is the famous *central limit theorem.* It applies to project situations provided that the number of activities is *reasonably* large. (For project management applications, four activities on the critical path are sufficient.) The theorem is independent of the probability density of each of the activities. (Indeed, each activity may be characterized by a different probability density.)

If we know the *expected time* t_i and the *variance* s_i^2 of each activity, A_i on the critical path, then the *expected duration* T_E of the project is

$$T_E = \sum_{i=1}^{J} t_i \qquad (4.5\text{-}1)$$

where there are J activities on the critical path. The *variance* S_E^2 *of the project* is

$$S_E^2 = \sum_{i=1}^{J} s_i^2 \qquad (4.5\text{-}2)$$

Let us define the project duration as D and define a parameter Z so that

$$Z = \frac{D - T_E}{S_E} \qquad (4.5\text{-}3)$$

The parameter Z is the number of standard deviations that the project duration exceeds the mean. The project duration D can then be found by using Table 4.5-1 for any given probability of project completion. For example, if the probability of completing a project is to be set to 90%, then, as shown in Table 4.5-1, $Z = 1.3$. Then, using Eq. (4.5-3), we see that the duration D_{90} corresponding to this 90% confidence is

Table 4.5-1 The Probability of Meeting the Completion Date Depends on the Number of Standard Deviations by Which the Proposed Project Duration Exeeds the Mean

Define: $Z = \dfrac{D - T_E}{S_E}$

Z	Probability of meeting completion date
3.0	0.999
2.8	0.997
2.6	0.995
2.4	0.992
2.2	0.986
2.0	0.977
1.8	0.964
1.6	0.945
1.4	0.919
1.2	0.885
1.0	0.841
0.8	0.788
0.6	0.726
0.4	0.655
0.2	0.579
0	0.5
−0.2	0.421
−0.4	0.345
−0.6	0.274
−0.8	0.212
−1.0	0.159
−1.2	0.115
−1.4	0.081
−1.6	0.055
−1.8	0.036
−2.0	0.023

$$Z = 1.3 = \frac{D_{90} - T_E}{S_E} \tag{4.5-4}$$

Thus

$$D_{90} = T_E + 1.3 S_E \tag{4.5-5}$$

The Gaussian probability density is shown in Figure 4.5-1, and the Gaussian distribution, which is obtained from Table 4.5-1, is shown in Figure 4.5-2. Using these figures, several obvious conclusions can be drawn:

1. If the duration of each activity in the project is given by its *expected* value, the project will be *late one-half* of the time. This is D_{50}, the estimate of the project's duration with a 50% confidence level.

2. Some engineers and technical specialists believe that they can build anything and that it will work immediately, that is, that testing is required only to verify that the activity does work. These engineers allocate too little time for testing, and when the activity fails to work, they fall behind schedule. We call this "optimistic" scheduling (or scheduling to be late!). If you schedule optimistically, the probability of completing a project on time has less chance of success than does winning at blackjack in Las Vegas.

3. If the risk and uncertainty in a project is small, the standard deviation S_E is much smaller than the expected time T_E, and we can, for all practical purposes, ignore S_E. In this case P-PERT is not needed.

4. During the lifetime of a project, the expected time t_i and variance s_i^2 of each activity is updated periodically, usually on a monthly basis. As the project progresses, the uncertainty inherent in each task will usually diminish, thereby decreasing s_i^2. When the project reaches its

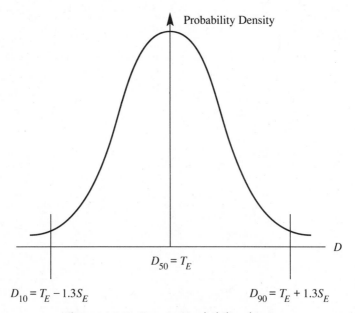

Figure 4.5-1 Gaussian probability density.

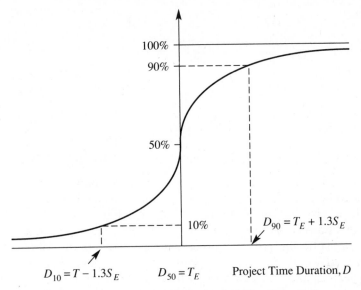

$$D_{10} = T - 1.3S_E \qquad D_{50} = T_E \qquad \text{Project Time Duration}, D$$

Probability of Completing the Project in a Time Interval, D

Figure 4.5-2 Gaussian distribution.

midpoint, it is usually possible to stop using P-PERT because S_E decreases sufficiently with respect to T_E and does not represent a significant impact on the schedule. At such a time, the standard precedence diagram is updated each month and uncertainty is neglected.

4.6 DETERMINING THE EXPECTED TIME AND VARIANCE OF AN ACTIVITY

The expected time t_i and variance s_i^2 of an activity A_i is readily found if the probability density of the activity duration is known.

4.6.1 Uniform Density

For example, if the probability density of the duration of an activity is known or assumed to be *uniform*

$$p(t) = \begin{cases} 1/(t_H - t_L); & t_L < t < t_H \\ O & ; \quad \text{elsewhere} \end{cases} \qquad (4.6\text{-}1)$$

then

$$t_i = \frac{t_H + t_L}{2} \tag{4.6-2a}$$

and

$$s_i^2 = \frac{t_H^3 + t_L^3}{3(t_H + t_L)} \tag{4.6-2b}$$

4.6.2 Statistics When First-Pass Duration is T_0 With Probability P

If the duration of an activity, such as fabricating an integrated-circuit (IC) chip, is to be characterized, we note that the chip can be fabricated in a time T_o with a probability of success P. If the chip is flawed, however, an additional time T_o is required. If once again the chip is flawed, an additional time T_o must be allowed. This process is repeated until the chip is fabricated properly. If we assume that the probability of success remains equal to P, the expected time t_i to successfully fabricate the chip is

$$t_i = T_o P + 2T_o P(1 - P) + 3T_o P(1 - P)^2 + \ldots \tag{4.6-3}$$

where $T_o P$ is the average time to complete the activity on the first try, $2T_o P(1 - P)$ represents the total time $2T_o$ times the probability of failure on the first time $(1 - P)$ and the probability of succeeding the second time P, and so on. Evaluating Eq. (4.6-3) yields

$$t_i = T_o/P \tag{4.6-4}$$

the second moment, $t^2 = (s_i)^2 + (t_i)^2$, is

$$t^2 = T_o^2 P + (2T_o)^2 P(1 - P) + (3T_o)^2 P(1 - P)^2 + \cdots \tag{4.6-5}$$

Evaluation of Eq. (4.6-5) yields

$$(s_i)^2 = \frac{T_o^2}{P^2} [1 - P] \tag{4.6-6}$$

Example 4.6-1

If $t_o = 1$ month and $P = 0.5$, then,

$$t_i = 2 \text{ months} \tag{4.6-7a}$$

and

$$s_i = \frac{T_o}{P} \sqrt{(1 - P)} = 2\sqrt{1/2} = 1.4 \text{ months} \tag{4.6-7b}$$

4.6.3 t_i and s_i When Activities Are Characterized by a Beta Distribution

If the probability density of an activity is not known a priori but the project specialists believe, in their professional opinion, that the duration of the activity will most often be between some optimistic value t_o and some pessimistic value t_p and will most likely be completed at a time t_m^* *then we usually assume* the probability density to be beta-distributed. In this case

$$t_i = \frac{t_p + 4t_m + t_o}{6} \tag{4.6-8a}$$

and

$$s_i = \frac{t_p - t_o}{6} \tag{4.6-8b}$$

In this text, and in practice, we characterize t_i and s_i using Eqs. (4.6-8) unless a priori information exists.

In the well-written humorous book *The Mythical Man-Month* [1] Brooks, the author observes that *All activities are on time until it comes time for testing,* that is, until we see that the activity does not perform as the specifications or statement of work requires. The author's remedy is to allocate sufficient time to an activity so that one-third of its estimated duration is for design, one-sixth for coding or fabrication, and one-half of the estimated duration left for preparing a test plan, testing, debugging and correcting, that is for making the activity conform to specification. If the estimated duration is T_A, the time allocated to design is then $T_A/3$, to fabrication or coding $T_A/6$ and to testing $T_A/2$. The estimated duration might be considered a *pessimistic* estimate.

The *optimistic* estimate assumes that the activity, when fabricated or coded, meets the specifications. In this case the time required for testing is seen as a "verification" of results and might be only take a time $T_A/10$. In this optimistic case $t_o = 0.6T_A$.

The expected duration and standard deviation of an activity (described as above), where $t_P = T_A$ and $t_o = 0.6\ T_A$, and where we assume, for simplicity, that $t_m = 0.8T_A$, is $t_i = 0.8T_A$ and

$$s_i = \frac{t_p - t_o}{6} = \frac{0.4T_A}{6} \approx 0.07T_A$$

---

* We recommend that t_m be selected by asking oneself whether the activity has a greater chance of being completed earlier or later than the midtime $T_m = (t_o + t_p)/2$ and choosing t_m to be less than or greater than $(t_o + t_p)/2$, respectively.

Many authors dealing with *quality* have coined the phrase *"Do it right the first time."* Yet we suggest that the time $T_A/2$ be set aside in a pessimistic schedule for *debugging* and *correcting*. Several of our students have asked whether the slogan and our pesimistic estimate are consistent. We would first observe that very few engineers ever build or code anything that meets specs without modification—this is the *reality* of the situation. *Do it right the first time* means that each unit, which forms the system to be built, should be built, tested, and made to meet specifications *before* the units are integrated into the system. Then during the system test phase, few "bugs" will be found.

Unfortunately, it often seems necessary, due to optimistic scheduling, to submit units for system integration before they are "bug free." "Don't worry," the engineer says, "we will find them in system test." In the system test, however, there are then so many bugs that they become extremely difficult to find.

Thus, *do it right the first time:* make each unit work according to specifications *before* system integration. This has been found to require a time allocation of T_A. Quality is a continuous process. It starts at the project's conception and continues throughout its duration. Quality takes time and this time must be allocated.

Example 4.6-2

Task Designation	Precedence	t_o	t_p	t_m
A	*Start of project*	12	16	14
B	Start of project	1	5	3
C	A, B	2	5	5
D	B	5	15	7
E	C, D	2	6	4
F	E	6	14	10

Table 4.4-1 is a work breakdown in which each proposed activity duration is given by its expected value. In this example, the duration of each activity is specified by the most probable time t_m, the optimistic time t_o, and the pessimistic time t_p.

  a. Determine the critical path.
  b. How long will it take to complete the project, with a 90% confidence?
  c. With what confidence can we complete the project in 31 months?

Solution

We begin our solution by using Eqs. (4.6-8) to calculate (t_i), (s_i), and (s_i^2) for each activity. The result is presented below:

Activity	Precedence	t_o	t_p	t_m	t_i	s_i	s_i^2
A	Start	12	16	14	14	2/3	0.44
B	Start	1	5	3	4	2/3	0.44
C	A, B	2	5	3	3.2	1/2	0.25
D	B	5	15	7	8	5/3	2.8
E	C, D	2	6	4	4	2/3	0.44
F	E	6	14	10	10	4/3	1.8

a. The precedence diagram for this example is shown in Figure 4.6-1. There are three paths through this network, *ACEF*, *BCEF*, and *BDEF*. The paths, their expected durations (T_E), and their variances (S_E^2) are given below:

Path	T_E	S_E^2	S_E
ACEF	31.2	2.93	1.7
BCEF	21.2	2.93	1.7
BDEF	26	5.48	2.3

If the critical path were defined in terms of the expected duration, the critical path would be *ACEF*. Now, however, we see that path *BDEF* should also be investigated, since its expected duration is 26 and its variance is *twice* the variance of path *ACEF*.

b. For a 90% confidence, the duration D_{90} of each *path* is

Path	D_{90}
ACEF	33.4
BCEF	23.4
BDEF	29

We therefore conclude that *ACEF* is critical and that there is a 90% probability of completing the project in 33.4 months.

c. Because the expected project duration is $t_E = 31.2$ months, the probability of completing the project in 31 months is somewhat less than 50%.

4.6.4 Calculation of the Critical Path When the Randomness of Activities Is Included

In order to determine the critical path, we could calculate the duration D_j of each path j, where

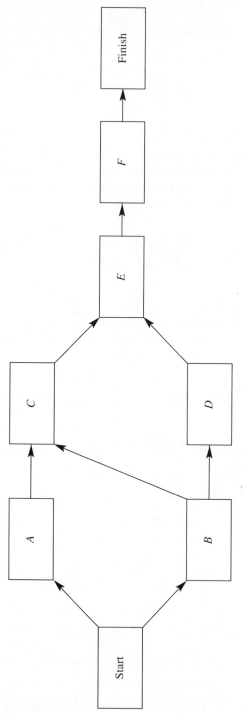

Figure 4.6-1 Precedence diagram for Example 4.6-2.

$$D_j = \sum_{i=1}^{N} t_{ji} + 1.3 \sum_{i=1}^{N} s_{ji}^2 \qquad (4.6\text{-}9)$$

where $i = 1, 2, \ldots, N$ are the activities on path j. The critical path is then the maximum of all the j paths:

$$D_{\text{critical path}} = \max \{D_j\} \qquad (4.6\text{-}10)$$

In a typical project there may be several thousand activities and the total number of paths could exceed 100. The problem of calculating the critical path using Eqs. (4.6-9) and (4.6-10) is seen to be extremely complicated. However, using the procedure of *dynamic programming* outlined in Section 4.4-2, the complexity reduces significantly. In this procedure the duration of each path starting from the *start* node is calculated up to each node where several paths intersect. Let us call one such node N_α. Following the procedures outlined in Section 4.4-2, we select the path having the *largest* duration. In the context of P-PERT this means selecting the maximum partial duration D_α where

$$\max \{D_\alpha\} = \max \left[\sum_{i=1}^{N_\alpha} t_{ji} + 1.3 \sum_{i=1}^{N_\alpha} s_{ji}^2 \right]$$

where t_{ji} and s_{ji} are the mean value and standard deviation of activity i on path j and node N_α is the node where each path $j = 1,2, \ldots, J$ intersect. The path chosen is the path which maximizes D_α. This technique is continued until N_y is the last node. At this point only one path remains, and with 90% confidence the project will last a time D_{90} where

$$D_{90} = \sum_{i=1}^{N_\lambda} t_{\lambda i} + 1.3 \sum_{i=1}^{N_\lambda} s_{\lambda i}^2$$

where $j = \lambda$ is the maximum duration path, i.e., the critical path.

Example 4.6-3

To illustrate our approach, we revisit Example 4.6-2. The precedence diagram for this example is shown in Figure 4.6-1. Note that at node C there are two partial paths: A and B. Their durations are:

$$D_A = 14 + 1.3(2/3) \approx 14.9$$

and

$$D_B = 4 + 1.3(2/3) \approx 4.9$$

Because D_A is the larger, path BC will *not* be on the critical path.

We also have two partial paths intersecting at node E. They are AC and BD. The duration of each path is

$$D_{AC} = 14 + 3.2 + 1.3 \sqrt{0.44 + 0.25} = 18.2$$

and

$$D_{BD} = 4 + 8 + 1.3 \sqrt{0.44 + 2.8} = 14.3$$

Hence path BDE is not on the critical path. Thus, the critical path is $ACEF$ and its duration is

$$D_{ACEF} = 14 + 3.2 + 4 + 10 + 1.3 \sqrt{0.44 + 0.25 + 0.44 + 1.8} = 33.4$$

as expected.

4.7 RISK AND UNCERTAINTY

Risk and uncertainty are directly related to the differences between the pessimistic and optimistic times t_p and t_o. Indeed, if there is no uncertainty and hence no risk involved in performing an activity, the activity is deterministic and $t_i = t_p = t_o = t_m$. The standard deviation is then $s_i = 0$.

Probabilistic-PERT analysis, in addition to showing us the longest or critical path (CP), indicates which activities on the critical path have *high* uncertainty or risk. This is evidenced by a large deviation s_i. For example, in project FWQ, an activity that (we shall call 1011) is on the critical path and has high risk because the assigned project personnel are inexperienced. There is another task, however, activity 1028, which does not lie on the critical path and has experienced personnel capable of doing the work called for on task 1011. It is then wise to consider exchanging these personnel. This exchange will decrease the risk and, therefore, the standard deviation of the CP and, hence, would decrease the project's duration.

It is often useful to describe the risk (uncertainty) of each activity in terms of the parameters L (low), M (medium), and H (high). A short, one-sentence or brief-paragraph description explaining your choice is also appropriate, especially when the risk factor is controversial. Suppose, for example, that a proposal you and your team are submitting to the government relies on a technology (activity 1026) that you and your colleagues have pioneered. You have patents and publications indicating that activity 1026 is low in risk because you, its creator, are performing it. On the other hand, to anyone else, including the proposal reviewer, the technique used to perform this activity is extremely high in risk. Your capability with regard to this activity should be explained. Similarly, assume that you choose to use a high-risk approach to complete another activity (1051) that is not on the critical path. Even if the activity's duration ends at t_p, the new technology may result in future benefits to your company. This rationale must be fully explained.

A simple guideline is:

Risk	$R = (t_p - t_o)/t_p$
Low	$R < 0.25$
Medium	$0.25 < R < 0.5$
High	$R > 0.5$

4.8 CRASHING THE PROJECT

Consider the possibility that after performing a probabilistic-PERT analysis you determine that the project duration will result in missing the due date set by the customer. One technique that can be employed to shorten or "crash" the project is to increase personnel in selected activities.

To reduce the project duration by increasing personnel makes one wonder, "If I doubled my personnel, would I halve its duration?" The answer is usually "NO!" Indeed, if we plan an activity to last T_A and use P_A personnel, a typical personnel-time plot would resemble that shown in Figure 4.8-1.

In Figure 4.8-1 we see that for a given activity A, the number of people needed to complete the activity in a given period of time is P_A . The duration

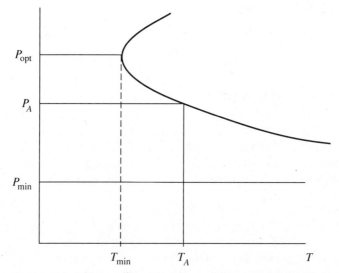

Figure 4.8-1 People trade-off for an activity, A, being performed.

of the activity can be decreased to T_{min} by increasing the number of people to its optimum level (P_{opt}). However, if more people than P_{opt} are employed for this activity, the activity duration actually *increases*. Thus T_{min} is the minimum possible time to complete task A. Indeed, if an activity can be *partitioned*, then one can shorten the activity by increasing the number of *appropriate* personnel. Even in this case however, increasing personnel increases the need for communication thereby increasing time. In the limit, too many people may actually increase the task duration.

If the activity cannot be partitioned, however, the use of additional personnel will result in time being wasted and actually increase the project duration. For example, consider that two expert chefs are involved in making a souffle. The addition of a third chef would probably serve to increase the time to make the souffle since there will be increased conversation and conflict.

On the other hand, if people are to be removed from an activity, we reach a point where the further reduction of personnel may result in the activity taking an *infinite* amount of time. For example, consider a construction site where several employees are used to move a steel beam. If all except one of the people are removed from the job, it may be impossible for one person to move the beam and accomplish the job. The duration of the activity, therefore, becomes infinite.

An activity can also be crashed by increasing the nonhuman resources. For example, if we have access to a large mainframe computer but, because of the backlog of potential users, will have to wait 3 weeks to run a job, we could shorten the activity's duration by purchasing a personal computer.

4.8.1 Which Tasks to Crash

Now that we have seen that it is possible to shorten an activity by increasing personnel or nonhuman resources, we must next determine which activities should be *crashed*. The answer is obtained in two parts:

1. Crash an activity only if it is on the critical path.
2. To minimize *cost,* crash the activity on the critical path that shortens the activity the most for a given expenditure. Thus, if two activities, A and B, are on the critical path and if adding one person to activity A will decrease its duration by 2 months but adding one person to activity B will decrease B by 3 months, then add the one person to activity B.

Example 4.8-1
The development of the satellite communication system connecting the New York branch of BTR Investments with their Dallas branch was late. The project had already cost $9.2 million. The Project Manager, Donna Kasher, knew that each month that the system was late meant a $50,000 loss due to expenses that would not occur once the system was completed. Donna's job

Table 4.8-1 Activity Crash Report

Activity	Precedence	Time Remaining (months)	Cost Remaining	Additional Cost/Month to Crash Activity	Number of Months That Activity Can Be Crashed
A	—	3	60,000	30,000	1
B	A	5	120,000	20,000	1
C	A	5	160,000	50,000	2
D	A	4	80,000	10,000	2
E	C, D	2	60,000	15,000	1
F	B, E	3	140,000	40,000	2
			$620,000		

was in jeopardy. She called an emergency meeting of the activity managers to determine what strategies would shorten their activities and hence the project. The report received is shown in Table 4.8-1.

The project was originally scheduled to be completed in 7 months.

a. Plot the cost of the project as a function of time.
b. What is the project duration if Donna is to complete the project with minimum cost?

Solution

Figure 4.8-2 represents the BIR Investments' possible system's precedence diagram. As indicated, the critical path, shown by the solid arrows, contains activities *ACEF* and, if nothing is done to crash the project, it will take an additional 13 months for completion. Hence, the project will be 6 months late and BTR would lose $300,000.

Naturally, Donna wishes to minimize this loss. After reviewing Table 4.8-1 and the precedence diagram, she plans to crash an activity on the critical path. Since *E* is the least costly, $15,000/month, Donna's first selection is to crash *E* from 2 months to 1 month. This would result in shortening the project by 1 month, thereby saving $50,000 less the $15,000 increased cost of crashing. In all, Donna would save $35,000.

After rechecking the precedence diagram, shown in Figure 4.8-3, the critical path is seen to be *ACEF*. Because *A* was the least costly to crash ($30,000/month), Donna now decides that *A* should be crashed by 1 month. The savings would then be $50,000 less $30,000, which is the increased cost incurred by crashing *A*. This represents a net savings of $20,000.

The precedence diagram obtained after crashing *A* is shown in Figure 4.8-4. The critical path remains *ACEF*. Activity *F* is next selected for crashing. It

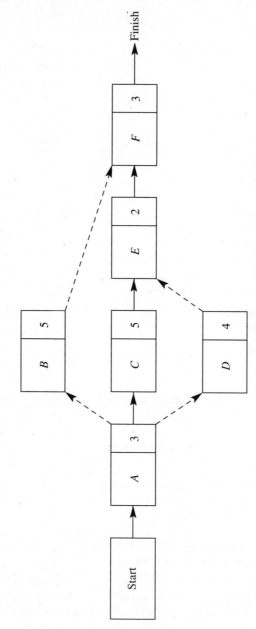

Figure 4.8-2 Precedence diagram for Example 4.8-1.

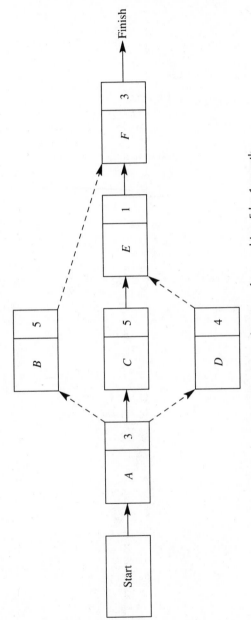

Figure 4.8-3 Precedence diagram after crashing *E* by 1 month.

123

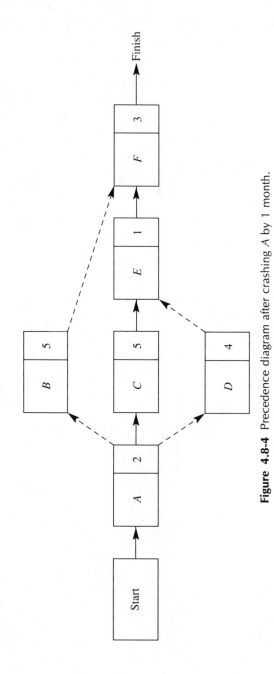

Figure 4.8-4 Precedence diagram after crashing A by 1 month.

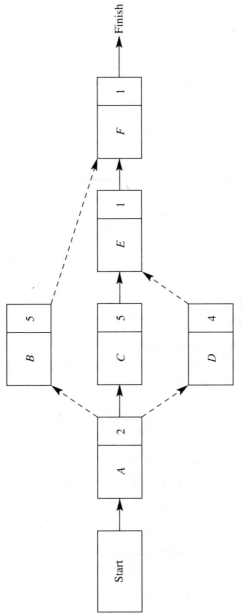

Figure 4.8-5 Precedence diagram after crashing *F* by 2 months.

can be crashed by 2 months and the project reduced by 2 months. The net savings is:

Gross savings by shortening project by 2 months	$100,000
Cost of crashing the project by 2 months	80,000
Net Savings	$ 20,000

The precedence diagram representing these crashing effects is shown in Figure 4.8-5. Note that *ACEF* remains as the critical path. Only activity *C* remains to be crashed. Crashing *C* by 1 month results in *no* net savings since we actually incur increased cost of $50,000 by crashing *C*. As illustrated in Figure 4.8-5, after crashing *C* by 1 month, from 5 months to 4 months, all three paths are now critical.

Keeping in line with her plan to reduce project costs, Donna decides to crash activity *C*. The precedence diagram obtained after crashing *C* by 1 month is shown in Figure 4.8-6. To crash the project further requires activities *B, C,* and *D* to be each reduced by 1 month. This results in a net loss. Crashing would involve the following costs:

Activity	Added cost
B	$20,000
C	50,000
D	10,000
Total increased cost due to crashing	$80,000

We would gain $50,000 and therefore *lose* $30,000 by crashing activities *B, C,* and *D* by an additional month.

The final cost–time curve is shown in Fig. 4.8-7. As a result of crashing the project, Donna selected to crash activities *ACE* each by 1 month and activity *F* by 2 months. In this way the project may be completed in 8 months. The total increase in cost due to project delays and crashing costs was $225,000, rather than the $300,000, which would have resulted from not crashing the project.

4.8.2 A Flow Diagram for Crashing a Project

Figure 4.8-8 showns the flow diagram used to crash a project. This flow diagram is readily programmed on almost any microcomputer.

The diagram begins with a table, such as Table 4.8-1, listing each activity, its normal duration, the additional cost–unit time incurred by crashing an activity, and the total time that an activity can be crashed. Next the critical path (or paths) is (are) found, the activities on the critical path determined, and the activity offering the most favorable cost–time trade-off located. The

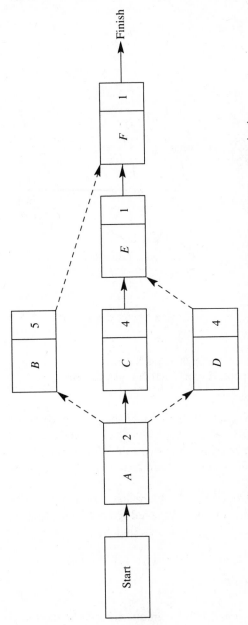

Figure 4.8-6 Precedence diagram for Figure 4.8-1 after ACEF are crashed.

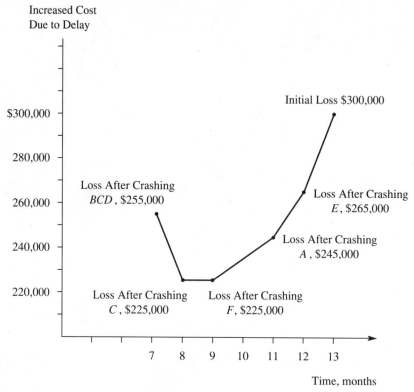

Figure 4.8-7 Cost-time curve showing that by crashing the project the increased project cost can be reduced from $300,000 to $225,000.

selected activity (or activities) is (are) then crashed by one unit of time.*
This process repeats until it is no longer possible to crash the project.

A cost–time curve (similar to Figure 4.8-7) is then plotted and a *business decision* can then be made regarding to what degree the project should be shortened.

If an R and D program is to be crashed, the inherent uncertainty in each activity may result in an increased project duration rather than in a decreased duration due to the fact that after crashing there are multiple critical paths.

4.9 GANTT CHART

The *Gantt,* or *bar chart,* is a technique for displaying the project schedule. Using the Gantt chart, the project engineer can see the early start–early

---

* A unit of time is the number of days, weeks, or months used in the scheduling process.

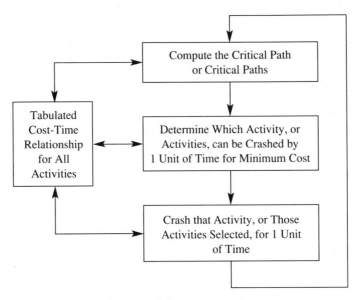

Figure 4.8-8 Flow diagram for crashing a project.

finish and the late start–late finish times of each activity or any subset of activities in the project. The Gantt chart further indicates progress and highlights the critical path. It does not, however, indicate precedence relationships, although even this feature has been included by certain Project Management software system manufacturers.

To illustrate the use of the Gantt chart, refer to Example 4.4-1. The precedence diagram for this example is shown in Figure 4.4-3, and the Gantt chart for this project is shown in Figure 4.9-1. Here, each activity is represented as an open solid bar (or rectangle) illustrating the activity's early start (ES) and early finish (EF) time. If the activity is on the critical path, a horizontal line is drawn through the middle of the bar, as seen in activity 100. Some software project management systems represent an activity on the critical path by using a different color, such as red, or by putting a *dashed* box around the solid rectangle. If the activity is not on the critical path, a dashed rectangle is used to show the late finish (LF) of an activity.

To determine the ES and EF times, we start at $T = 0$, whereas to determine the LS and LF times, we start at the end of the project, which is month 46. To illustrate, consider tasks 400 and 410. Figure 4.4-3 indicates that task 400 starts after task 200 ends. This represents the ES time. The EF time is 3 months later. To find the LF time of task 410, examine Figure 4.4-3. Note that task 410 must finish before task 500 begins. Since 500 is on the critical path, 410 must end no later than month 36. Thus, the LF of activity 410 is month 36 and the EF of activity 410 is 3 months earlier at month 33. The LF of activity 400 is then at month 33 and its EF is month 30. The *slack* associated with an activity is the time interval between the ES and the EF.

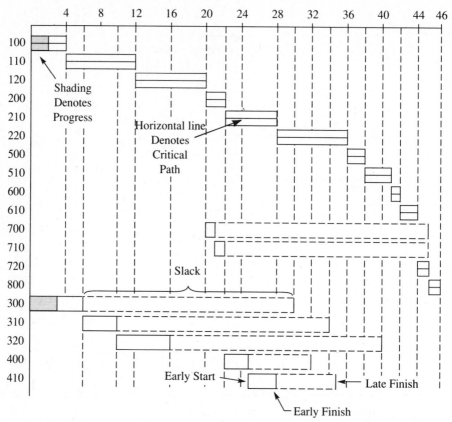

Figure 4.9-1 Gantt chart for Figure 4.4-3 after 4 months have elapsed.

The Gantt chart is updated and redrawn every week or month, depending on the length of the project. In this manner, project progress (positive or negative) can be easily reported. This progress is usually shown by shading in a portion of each activity according to the percentage of work completed. For example, if at the end of month 4, only 50% of activity 100 and 50% of activity 300 are completed, the Gantt chart would appear as in Figure 4.9-1. Since we have completed month 4, we are *behind* on our critical path and if conditions do not improve, we will delay the project by 2 months.

4.10 PERSONNEL SCHEDULING

The Gantt chart is a useful tool to demonstrate how to schedule personnel. To illustrate, consider that in Figure 4.10-1(*a*) all the personnel listed above each activity bar are software development experts who can work on any of the activities illustrated with equal capability. Figure 4.10-1(*b*) shows the

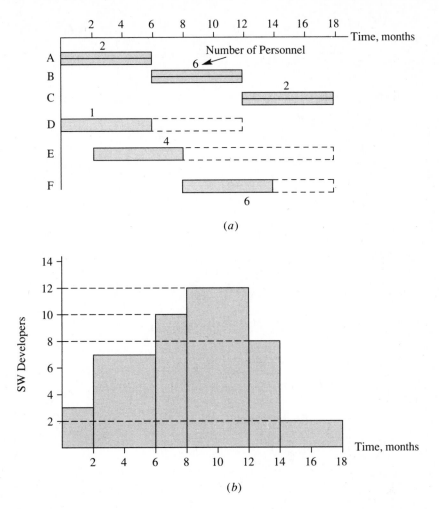

Figure 4.10-1 (a) Gantt chart showing the number of software developers for each activity; (b) software development personnel used as a function of time; (c) cumulative distribution of software development personnel as a function of time. (*continued*)

sum of the personnel used in each task per month and how many software development specialists are needed as a function of time, when each activity is started as early as possible. Figure 4.10-1(c) shows the *cumulative distribution* of software development personnel as a function of time. This curve indicates the total number of staff–months that we plan to use as a function of time and is extremely important when planning a *budget*, since worker–months are directly related to *dollars*.

For example, if software development specialists earn a *loaded* salary of $150,000/year or $12,500/month, then after 6 months the project is expected

Personnel – Months Employed

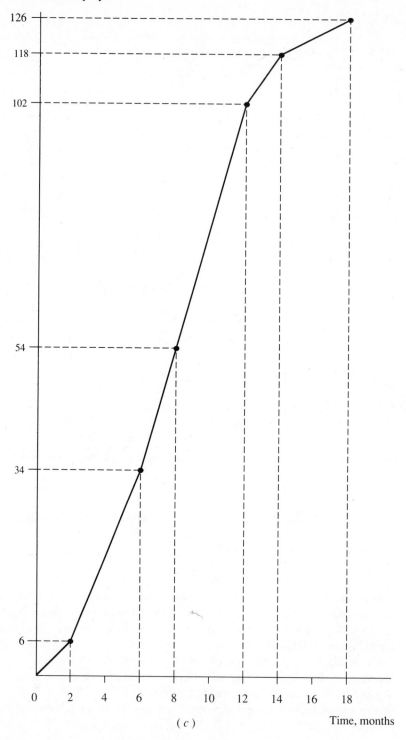

(c)

Figure 4.10-1 (Continued)

to spend [see Figure 4.10-1(*c*)], 34 × \$12,500 =\$425,000. After completion, we expect the project to spend, 126 × \$12,500 = \$1,575,000 [see Figure 4.10-1(*c*)].

4.10.1 Leveling

Referring to Figure 4.10-1(*b*), we see that on the average, eight software development specialists are required for this project. However, between months 8 and 12, 12 such specialists are needed, representing a *peak* period for personnel use. It is often the case that we do not have sufficient personnel available to satisfy this peak value. To determine whether we can *minimize the peak* number of software development specialists, we move activities *D, E,* and *F, which are not on the critical path, from their ES positions to start times that will minimize the peak number of personnel required.* Figure 4.10-2 shows an alternate personnel plan, obtained by starting activity *F* at month 12, in which the peak number of software personnel has been reduced to 10.

If the number of personnel is critical, and time is not critical (such is often the case in a low-priority project), it is possible to decrease the number of specialists required by delaying the project. For example, if we have six software specialists available for the project illustrated in Figures 4.10-1 and 4.10-2 and additional personnel having these skills cannot be hired, then by delaying activities *C, E,* and *F,* we can reduce the personnel to six. Such a

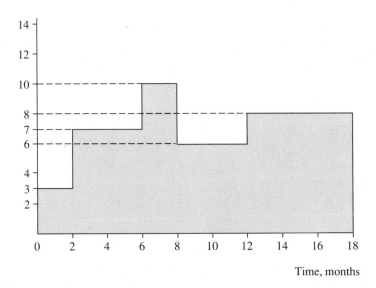

Figure 4.10-2 Late-start solution in which only 10 workers are needed.

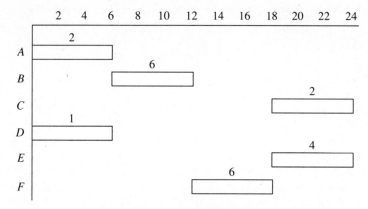

Figure 4.10-3 Gantt chart in which software development personnel is limited to 6.

solution is shown in Figure 4.10-3. Note that now the project takes 24 months to complete, representing an increase of 6 months.

Although the preceding examples were done by hand, *an actual project would contain more than just a single resource and six activities.* In such actual situations, software is available to plot personnel schedules, "level" them and allow what-if questions to be answered.

4.11 CONCLUSION

After the project is planned, that is, the WBS and work packages finalized; the Gantt chart, critical path, and P-PERT analysis performed; and personnel scheduled finalized, the project can be started. Chapter 5 describes how to *control* a project once it has begun.

REFERENCES

1. Brooks, Jr., F. B. *Mythical Man Month*. California:Addison-Wesley, 1982.

EXERCISES

1. Consider taking a trip to visit a relative in another city.
   a. If you were to take a train how long would the trip take? Give an "off-the-cuff" answer!
   b. Now lets estimate the trip duration more carefully. Estimate the time to go from your house to the train (assume its rush hour) by taxi.

Estimate the time to walk to the ticket counter from the taxi, to purchase a ticket, to wait for the train, for the train ride, to wait for a taxi at your destination and finally the time to travel by taxi to the relative's house.

   c. Compare (a) and (b). Usually (b) greatly exceeds (a). If your original estimate did not exceed (b), your estimate was either based on experience, or you performed (b) without thinking about it!

2. In Fig. 4.4.2 assume that activity A, has been completed. No other project activities have been started. Find the critical path and the project duration.

3. Refer to Example 4.4-1. Assume that Tasks 100, 110 and 300 have been completed, and Task 120 has 2 months work remaining. No other project tasks have been initiated.

   a. Determine the critical path.

   b. Calculate the time remaining in the project.

4. Redraw Fig. 4.4-3 using the *activity-on-arrow* technique.

5.

JOHN GOES TO SCHOOL

		Time	Precedence
1.0	*Mr. Ready*		
1.1	Awakens to alarm	Start	
1.2	Awakens wife	1	1.1
1.3	Awakens John	4	1.2
1.4	Brushes teeth	1	1.3
1.5	Showers	5	1.4, 2.2
1.6	Shaves	5	1.5
1.7	Selects clothing	2	1.6
1.8	Dresses	5	1.7
1.9	Walks dog	10	1.8
1.10	Eats breakfast	10	1.9, 2.6
1.11	Gets into car	1	1.10
1.12	Takes train to work	60	1.11, 2.9
2.0	*Mrs. Ready*		
2.1	Wakes up	Start	1.2
2.2	Showers	10	1.2
2.3	Brushes teeth and puts on makeup	20	2.2
2.4	Dresses	10	2.3
2.5	Feeds dog	2	2.4
2.6	Makes breakfast	5	2.5
2.7	Eats breakfast	10	2.6
2.8	Prepares children's lunch	10	2.6

continued

JOHN GOES TO SCHOOL (CONT)

		Time	Precedence
2.9	Drives children and Mr. Ready to station	10	2.8, 1.11, 3.6
2.10	Continues on and takes John to school	15	2.9
2.11	Drives to work	20	2.11
3.0	*Ready Children*		
3.1	Awaken and get out of bed	2	1.3
3.2	Shower	15	2.2, 3.1
3.3	Brush teeth	1	3.2
3.4	Dress	5	3.3
3.5	Eat breakfast	5	3.4, 2.6
3.6	Get into car	1	3.5
3.7	Arrive at school	End	3.6, 2.10

Use the above information to solve the following:

a. Draw a precedence diagram.

b. Determine the critical path.

c. Draw a bar chart using early starts.

d. How long does it take the Ready children to arrive at school?

e. What events lie along the critical path?

f. How would you shorten tasks, rearrange precedence relationships or add resources so that Mr. and Mrs. Ready can sleep later in the morning?

g. How many bathrooms are needed if Mr., Mrs., and the Ready Children do not simultaneously use a bathroom (the children can share a bathroom) and early starts are used.

h. (*a*) Is it possible to have Mr. Ready, Mrs. Ready and the Ready children use the same bathroom at different times without Mr. or Mrs. Ready or their children awakening at an earlier time? (*b*) If it is not possible, what is the latest time that either Mr. Ready, Mrs. Ready, or their children can awaken.

6. An integrated circuit is to be fabricated. The time duration of each pass is 60 days. On the first pass, the probability of success is 30%, on the second pass 60% and on the third and subsequent passes 90%. Estimate the expected duration t_i and the variance s_i^2.

7. Verify Eq. (4.6-6).

8. If $T_o = 90$ days and $P = 0.9$, calculate t_i and s_i.

9. Repeat Problem 1b. Assume an optimistic, pessimistic, and most likely duration for each step in the trip. With 90% confidence how long will the trip last?

10. Refer to Example 4.4-1. The listed duration is t_m, t_o, t_m, and t_p are given below.

a. Find the critical path

b. With 90% confidence, how long will the project last?

c. With what confidence-level (probability) will the project last no longer than 49 months?

Task	t_o	t_m	t_p
100	3	4	6
110	6	8	10
120	6	8	16
200	2	2	2
210	4	6	8
220	6	8	10
300	3	6	9
310	2	4	8
320	5	6	15
400	2	3	4
410	3	3	3
500	2	2	2
510	2	3	6
600	1	1	1
610	2	2	2
700	1	1	1
710	1	1	1
720	1	1	1
800	1	1	1

11. Repeat Example 4.8-1 if the precedence relationships for each activity is changed to that given below. All other information is the same. Obtain a "crash" curve similar to Fig. 4.8-7.

Activity	**Precedence**
A	—
B	A
C	A
D	B
E	C,D
F	B,E

12. Verify Fig. 4.10-3.

CHAPTER 5

CONTROLLING THE PROJECT

5.1 INTRODUCTION

All projects must be rigorously and constantly monitored at regular intervals throughout their life cycles to assure compliance with technical performance standards and targeted fiscal objectives. To measure how successful the project is at achieving engineering directives and satisfying market requirements, businesses exercise control over the quality, schedule, and cost of work to be performed. Daily operations are monitored through formal and informal reporting procedures to determine how well projects are progressing and whether initial goals and objectives will be met.

This chapter discusses, in detail, appropriate control techniques used, by government and commercial R&D Project Managers, to manage successfully ongoing project performance.

5.2 THE ELEMENTS OF PROJECT CONTROL—THE PROCESS OF CONTROLLING A PROJECT

The project controlling process is an *evaluative* process whereby deviations from planned events are reported and probable causes assessed. It is a *performance measurement* process by which corrective action is taken to alleviate the impact of these deviations and other unfavorable trends on project schedules, budgets, resources, or staffing levels. Project control is also a *quality assurance* process intended to maintain the technical performance standards of the product under development and to assure the quality of the product design.

At the project level, management control is operationally defined because

TABLE 5.2-1   Steps in Establishing Project Control

1. *Establish a measurement system* that can compare what has been planned to what actually will takes place; this requires the setting of hard milestones and the development of test plans
2. *Measure results* and assess deviations from the original project plan:
   a. Schedule
   b. Budget
   c. Technical specifications
   d. Resource requirements (staff, materials, equipment, etc.)
3. *Report results* to the appropriate project personnel and managers
4. *Forecast the results of any deviation;* evaluate potential hazards; review trends to analyze their impact on project progress; discuss trade-offs
5. *Take corrective action* to bring about the originally desired goal and accomplish the initial project objectives; if necessary, one should replan

the monitoring and analysis processes, once put into place, allow the Project Manager to determine how well a project is progressing against planned objectives. In project oriented environments, the need to direct complex, interrelated organizations and facilitate continuous free flow of information pertaining to the project across organizational entities necessitates step-by-step analyses and decision-making. Therefore, the aim of the control process is to get project work done on time and within budget constraints.

Project control helps keep top management appraised of project status, specifically when slippages or budget overruns jeopardize contractual agreements, push out a project's end date, and thereby delay entry of the product into the marketplace. The functional executive's review of project control documents helps to clarify for all team members new project directives or corrective action that might be taken in light of other corporate priorities and strategic plans. Also, chief executives must constantly evaluate overall project status and thus want timely, formal reports to determine whether and when expenditures for a nondiscernable effort or project should be terminated.

Table 5.2-1 summarizes the project controlling process by listing its five fundamental steps.

Step 1, *establishing a measurement system,* begins in the planning phase, incorporating all the specifications needed to be achieved. It includes the setting of hard milestones and the development of *test plans* that are prepared by the engineers involved. Some test plans are approved by quality assurance personnel and agreed to by the customer. The test plan should (1) explain the specification that the test is to verify, (2) specify how the test is to be performed, (3) list what equipment will be required, and (4) describe the expected results.

In order to properly control a project, additional testing is also required to determine deviations from schedule, budget, and personnel. Too often, the

Project Manager is presented a status report to read that says "this is a 6-month effort, of which 2 months are now completed, we are on time, *there is no problem, trust me!*" Unfortunately, at the end of the sixth month we begin to see the problems arise: "We have found a minor bug that I am sure we will have no difficulty fixing." The result is often a several-month delay. We recommend that a *hard* test be planned *in advance,* for each month, for each activity. On some shorter activities perhaps biweekly tests are in order. In this way, the Project Manager can, indeed, *trust* the comments made in the status reports and, in fact, believe that there is *no problem.*

The *control* process is a *feedback process* as shown in Figure 5.2-1 in which deviations from the plan are reduced. In any control system, increased delay results in an increased tendency for the control system to become *unstable.* For the control process to operate properly, carefully designed tests, conducted in a timely manner, must be incorporated into the process. These tests are often accentuated by calling them *milestones.*

Step 2 in Table 5.2-1 involves *measurement of results* and *assessment of deviations* from the plan. It is very important to note that the tests performed are very specific and designed to meet specifications and assess progress. One should *not* design and perform a test that incorporates a specification not provided by the customer or by the company's *standard operating practice.* Likewise, the measurement and assessment should be performed not only by the one performing the work but also by the one who is the successor of the activity to ensure that both parties agree on the work's completion. In addition, the activity being tested cannot be judged to be completed unless *all* of the specifications relating to that activity have been met. It is only in this way that quality can be assured.

Deviations from the plan can be readily assessed after measurements are taken, and in Step 3, the results are reported in writing and orally to the appropriate personnel. We recommend a one-page status report designed to present the extent of any deviation in a clear, concise manner. We have found that engineers often try to "fix" the problem before submitting the report to make the report "look better." Such a delay in reporting could be catastrophic, causing instability in the control process. Long reports also delay reporting. Too often we hear the comment "Of course I'm late, I had to write a 20-page report."

In Step 4, the engineer, the manager, and all other appropriate personnel evaluate the results of the report. Trends are reviewed and trade-offs studied. Of course, caution is taken to avoid responding to each deviation by taking some corrective action. Indeed, unless there is some trend, corrective action should not be taken. Once again we refer to a physical control system. If sufficient frictional damping or resistance is not applied, large overshoots in performance may occur that may well extend the project duration rather than shorten it.

Step 5 is designed to carefully apply corrective action. We recommend establishing a *Change Control Board* made up of selective management and

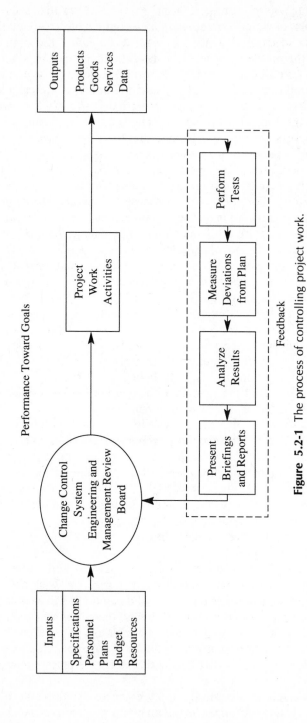

Figure 5.2-1 The process of controlling project work.

engineers to consider carefully any deviation and the impact imposed by the proposed change.

5.2.1 Guidelines for Establishing Project Control Systems

Meaningful control systems economically measure project information. They are *discernable* and *appropriate* for the complexity of the tasks being assessed and the size of the project effort. They are *timely, simple* to employ, and *congruent* with the events being measured [1].

1. In general, it is not necessary to collect data the project team doesn't intend to analyze. Likewise, the effectiveness of a project control system is not increased by using a computerized reporting program that generates huge amounts of statistical information. Nor are unread reports improved sheerly by their weight or volume. It is important to determine the minimum information requirements needed to analyze project situations to redirect or apportion work efforts prior to setting up the project control system. Capacity of the system established needs only to match these minimum requirements.

2. Trivia should never be measured. Control is maintained by assessing primary activities that affect overall performance and results.

3. Control any changes in the scope of project activity. Measure and compare results to predetermined standards; report only significant deviations and trends away from established plans. Remember, because a phenomenon can be measured, it does not mean that there is relevance in reporting it.

4. Controls must also be appropriate to the size and complexity of the activities or tasks being measured. In general, the more visible and larger the commitment of resources, the more managerial attention associated with the project and the greater the amount of control expended. However, technical performance, cost, and schedule objectives can go adrift in small as well as large and complex projects [2].

5. Measurement systems should be congruent with the events being measured. A measurement isn't more accurate if calculated to the Nth decimal place when it is at best a trend perceived in a diagram, chart, or graph.

6. Controls should be simple to employ. Costly systems may be too complex and may not satisfy the objective of providing the minimum amount of information to the Project Manager when needed most for interpretation and decision-making. Remember, to ensure that a control system is employed it must be *useful* and *useable*.

7. The control system must provide reports and generate data on a timely basis so that adequate corrective action can be taken before deviations from baselined plans become serious. Two dangers exist—measuring

Table 5.2-2   Uses and Characteristics of Modern Project Control Systems

System	Characteristics	Action Items Reporting	Budget and Resource Baseline Reporting	Breadboard Development	Change Control	Data Analysis	Documents Tracking	Performance Measurement	Prototype Development	Unit and Systems Test
Bar graph	Simple format, broad-based planning and tracking tool that is used for briefing and reporting	X	X	X		X	X	X	X	X
Configuration or change management procedures	Manual or computerized tracking system used to detect problems, modify requirements, and control project documentation				X	X	X			
C/SCSC-related reports	DOD cost/schedule performance reports required on all major U.S. government contracts		X			X		X		
Design reviews, code inspections, and walkthroughs	Technical assessments of product's integrity and compliance to design specs held by nonteam experts at key decision points			X		X			X	
Demonstrations	Quality assurance or end-user review of the working features of a software system to control product's reliability					X			X	X
Development statistics	Technical performance measurement statistics used to track progress, control quality, and predict completion date			X		X		X	X	X

Technique	Description							
Earned Value Analysis	Compares the budgeted cost of work scheduled, the budgeted cost of work performed, and the actual cost of work performed		X	X		X	X	X
Gantt chart	Simple format, similar to bar graph, time-phased planning and tracking tool that is used for briefing & reporting; good predictor of project completion date		X		X		X	X
Probabilistic PERT	Each network activity is figured as 3 separate time estimates; used where time uncertainty is a factor; a good predictor of project completion date		X			X		X
Project review meetings	Periodically scheduled business meetings of all project officials to review progress, identify problems, and solicit recommendations for improvements	X		X	X	X	X	X
Summary reports	Simple format, broad-based account of project status prepared by the project manager to brief upper-management, document changes, and report progress			X		X	X	X

too frequently so that reports are meaningless and extraordinarily expensive and not measuring often enough to effectively monitor the "pulse" of project activity. Managers must think through when and in what dimension (cost, schedule, technical performance) they must obtain results for successful achievement of project objectives. This is especially complicated in R&D projects that must culminate years later as prosperous commercial ventures.

8. Operationalize control systems. "Fit" them into the project organization. Give access to those members of the project team who need to interpret the data and who are capable of taking corrective action regarding the problem situations. Tailor the information to these people's needs. Reduce prose to graphics, and summarize project status in brief one- or two-page reports that are easy to read yet comprehensive. Remember, establishing an effective control system also requires paying attention to the interpersonal communication subsystems that exist within the project, as well as to the roles and authority of the various project leaders and managers engaged in the exchange of project information. Chapter 8 is devoted to this topic, as its importance to the establishment of meaningful and effective project control systems cannot be emphasized enough.

5.2.2 The Environmental Nature of Modern Project Control Techniques

There are several different measures or techniques that Project Managers must become familiar with in order to exercise proper control over project costs, schedules, and technical work. Table 5.2-2 describes modern project control techniques that are explained in this text. The table also illustrates where and when to implement each technique throughout the project to maintain control of R&D activities ranging in scope from single-team, fairly concrete development projects, to highly complex or uncertain research efforts. Just how effective a chosen control system is at measuring deviations in project performance against the initial estimates or baselined plans will depend on several variables. Moreover, since successful implementation varies most frequently on situational conditions present in the project environment, prescriptive models that might be used to select a control system are far from ideal.

Research conducted in an empirical study of 103 development projects sponsored by diverse industrial firms, NASA, and the DOD tested this relationship between situational conditions and how control systems were implemented [3]. Summarized below are the important conclusions reached by the author that have implications on how project managers might maximize the effectiveness of the control techniques they do select.

Control systems are interrelated to the project environment and are supported by:

- The degree of detailed technical planning performed by the project manager and the team leaders assigned to the project.
- The administrative prowess of the team management.
- The flexibility and consistency of project leadership over each of the life-cycle phases of the project.
- Careful analysis of the project environment prior to establishing control systems.
- Proper integration of the project schedule, budget, and technical tasks across each organizational and corporate entity contributing to the project work.
- Clear definition of reporting procedures to monitor or track the flow of project information.
- Identification and publication of all interface events, project sign-offs between phases, and engineering design changes approved by a Configuration Management or Change Control Review Board.

Specific techniques shown to be effective for monitoring and interpreting the performance of the project schedule in this study included:

1. Detailed interface planning, development of a WBS and a tracking control system, establishment of milestones.
2. Scheduled design reviews.
3. Milestone and Gantt chart reporting.
4. Network planning and control techniques using precedence diagrams.

A close interpretation of the study reveals that the more uncertainty associated with the scheduled and budgeted tasks, the closer a Project Manager must rely on control techniques that monitor technical performance. Control systems that emphasized only the tracking of project costs and schedules in lieu of formal design reviews had little effect on whether the project met design specifications. This was also true when product design was fairly complex. Yet technical monitoring of the devlopment process alone failed to yield technical success. Rather, it was shown that the careful integration and implementation of the techniques listed above, chosen to match the dynamics of the project environment, were responsible for the greatest impact on overall project success.

5.3 PROJECT COST CONTROL

As the engineering project proceeds through development, the Project Manager soon begins to realize the accuracy and true cost of the scheduled work in terms of capital outlays, resources, equipment, material supplies,

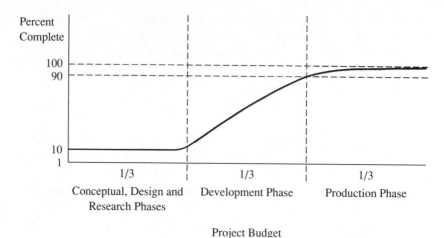

Project Budget

Figure 5.3-1 Plot of percent of work completed versus dollars in effort spent.

and other technical requirements. Because of the risky nature of R&D activity in high-tech industries and the scarcity of qualified professionals available to complete project assignments, management must monitor, from the onset of the project, both its technical and cost performance, as well as determine whether it is proceeding on schedule.

Let's expand on the concept of the project life cycle discussed in Chapter 3. Typically, early project expenditures are high when compared to actual performance of the scheduled activities. During the introductory or start-up phases, approximately one-third of the allotted budget is spent to complete 10% of the preliminary proposal, concept development, design feasibility, and engineering activities.

Engineering design and materials procurement employ another third of the project funds to accomplish 80% of the work during the middle phases. The remaining third is used during manufacturing and production to complete the remaining 10% of project work. Notice the slope of the plot shown in Figure 5.3-1. It illustrates how the cumulative distribution of the project budget versus the percentage of work completed approximates a "lazy S" curve. If the rate of work does not progress as planned, the curve will reflect delays in the planned activity and serve to alert the Project Manager to take corrective action (see Figure 5.3-2). An analysis of curves such as those shown in Figure 5.3-2 is called the *Earned Value Analysis* (EVA) and is discussed in detail in Section 5.3.2.

Once the project gets under way, the Project Manager must correlate the control of project schedules and costs. A summary of such measures includes:

- Periodic reestimation of the time and cost-to-complete remaining work.

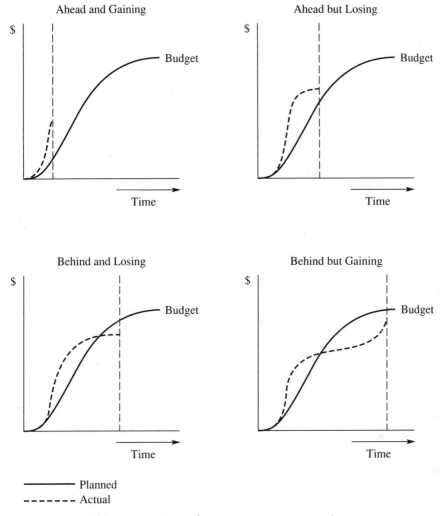

Figure 5.3-2 Typical project rate curves over time.

- Timely measurement of physical progress and comparison to planned and actual expenditures.
- Frequent and periodic comparison of actual progress and expenditures measured at both the activity and project completion levels.
- Disciplined budget and authorization of expenditures.

Of course, a well-organized schedule and cost control system must be planned for at the time that project goals and objectives are set. The integration of project schedule and cost control is readily achieved by correlating:

- Cost-coded, WBS element descriptors.
- Cost account and/or organization codes.
- Corporate financial ledger accounts.

Determining project costs as a function of time requires the summarization of project activity in dollars or hours of work performed—a task easily tracked by computerized project management software systems that also generate time-phased network plots of the project's schedule.

The major reason for establishing an integrated schedule and cost control system is for the Project Manager to obtain sufficient detail about task performance to forecast new estimates of how long it will take to complete the project and, at the same time, adjust and/or revise the schedule of these activities. Such control mechanisms allow management the flexibility to reallocate resources to prevent unwanted delays or premature termination of promising projects as a result of inadequate funding.

5.3.1 Causes of Cost Problems

When a project budget begins to overrun, the first steps the Project Manager should take are to (1) measure the overrun and (2) understand the probable causes of the overrun before attempting to remedy it.

The following is a list of factors that cause cost escalation on small and large projects:

- Tendency to base cost estimates for project resources and/or capital outlays on vague feasibility studies taken from similar project efforts rather than detailed design specifications or system requirements of the technology under development.
- Failure to build slack into the project schedule to realize the sought economies from cutting project work or product deliverables (features) in an effort to alleviate cost overrun.
- Delay in the flow of information on which to base early corrective action.
- Lack of corporate structures or unforgiving management styles that do not support open and honest disclosure of information that impacts or jeopardizes the project's financial success.
- Failure to modify, redesign, or reengineer the project when design flaws are first discovered or technical performance capacity falls below design objectives.
- Resistance to operationalize quality improvement efforts that are at odds with initial project cost goals, technical objectives, or performance specifications.
- Failure to recost objectively the entire project after modifications or

reengineering is authorized; the cost of added rework also must be measured.

- Overutilization of computer hardware and staff additions to late software development projects that drive up the operating costs of debugging the software.
- Incorporation of changes in design scope to the project contract without client or management approval, nor the appropriate allocation of funds for the changes.
- Indiscriminate distribution of the project's contingency budget to functional management who have authorized design changes that surpass expected limitations.
- Failure to adequately estimate the effects of inflation on total project costs—R&D, materials, component parts containing precious metals, labor, pilot plants, etc.

5.3.2 Earned Value Analysis

Earned Value Analysis, a technique used by both Project Managers and Accountants to track cost deviations and progress in a project, compares the prepared cost estimate, or budget, with the expenses actually incurred throughout the project. These comparisons usually are made monthly or biweekly. Further, EVA compares the cumulative actual costs, incurred up to a given date, with the estimated costs of project work actually completed for any specified period of time. Known as the project's *earned value,* these cost and schedule deviations enable project personnel to estimate the final cost of the project and to obtain the final completion date.

Cost Estimation (Budgeting). Budgeting R&D projects is a complex process requiring input from corporate financial planners and technical managers alike. Good budgets link the annual, near-term planning of the technical organizations to that of the business's overall, long-range strategic plans. No matter which management cost and control system is implemented to allocate scarce resources (incremental-, zero-, variable-based, or planning-programming budgeting), the objective is to develop a plan against which a time-phased budget can be established.

Budget preparation begins with the WBS. As we saw in Chapter 3, the work package associated with each lowest-level activity must describe the personnel needed to perform a given activity and the work schedule. Figure 5.3-3(*a*) shows a section of a WBS containing activities 111, 112, and 113 that together make up activity 110. Figure 5.3-3(*b*) shows the personnel needed on each of the activities, the duration of time that each is needed, and the loaded cost (i.e., the direct labor cost plus overhead and general and administrative expenses). Profit is often added later. Figure 5.3-3(*c*) shows each activity, when it occurs, its duration, and the number and type of

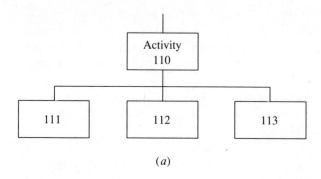

(a)

Activity ID	Skill Time (person-months)		Loaded Cost ($/month/ person)
111	1 (p-m)	Technician	$2,250
	1 (p-m)	Programmer	$2,300
112	2 (p-m)	Programmer	$2,300
	1 (p-m)	Engineer	$3,000
113	2 (p-m)	Programmer	$2,300
	3 (p-m)	Engineer	$3,000

(b)

(c)

Figure 5.3-3 (a) Part of WBS showing bottom-line activities 111, 112, and 113; (b) work breakdown structure activity data; (c) bar chart indicating number and type of personnel required as a function of time.

Incremental Personnel Cost for Month	$4,550	$5,300	$7,600	$5,300	$3,000	$5,900
Cumulative Personnel Cost for Month	$4,550	$9,850	$17,450	$22,750	$25,750	$31,650

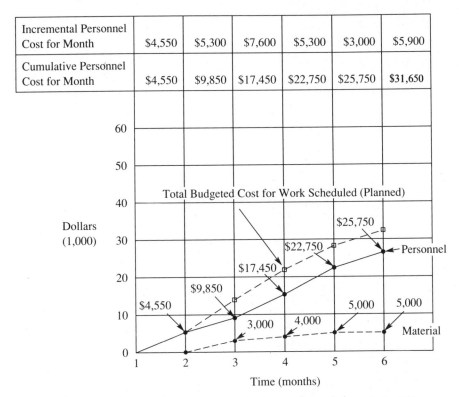

Figure 5.3-4 Budgeted cost of work scheduled (planned) for activity 110.

personnel required. Using Figure 5.3-3, one can plot the estimated cost of activity 110 as a function of time. (Note that activity 110 is the sum of activities 111, 112, and 113). This is shown in Figure 5.3-4. Also shown in this figure is the cost of materials for activity 110. Thus we arrive at the total *Budgeted Cost for the Work Scheduled* (BCWS). Since this cost curve was prepared in the planning phase, it represents the estimate against which actual expenses are compared.

The cost of all the other activities are also calculated and one final BCWS is plotted for the project. A typical resulting BCWS is shown in Figure 5.3-5. In this figure, we note that the total planned project cost is the *Budget At Completion* (BAC). Similarly, the planned *Time At Completion* is denoted TAC.

Actual Cost of Work Performed (ACWP). Each month, as work is being performed, each employee should complete a time sheet indicating how much time has been allocated to each of the many project tasks. Time spent marketing, proposal writing, or performing other activities not related specifically to the project should be charged to another charging number and

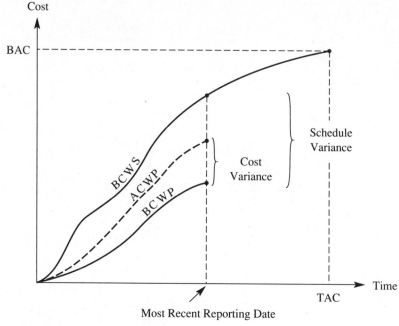

Figure 5.3-5 Earned value.

not to the project. Many companies require daily breakdowns by the hour, and this breakdown should be done as carefully as possible by each person on the project. It is estimated that one spends no more than 4 hours of an 8-hour day performing tasks associated with the activities they might be charged to, in that the remaining time is spent conversing on the telephone, socializing, reading, giving and asking advice on a different project, seeing customers about new business, or attending other unrelated meetings.

For example, all personnel working on activities 111, 112, and 113 file ACWP time sheets. Individual totals are summed and the results added for each major activity. Finally, one single ACWP is presented as illustrated in Figure 5.3-5.

Note that in Figure 5.3-5 the ACWP shows the project's actual spending and that it has currently spent less money than expected. Yet how much "work" has actually been accomplished? For example, if no work was accomplished at the time of the last report, even though money was spent on labor and material, then it could be argued that the project is "over budget." However, if the project is completed without spending all the allocated funds, then it is argued that the project is "under budget"—a nice position to be in!

Budgeted Cost of Work Performed (BCWP). The *Budgeted Cost of Work Performed* (BCWP) is a technique employed to indicate the budgeted or planned cost of the work "actually performed."

The curve shown in Figure 5.3-5 is often difficult to obtain. Without good communication, strong leadership, and carefully designed test plans, Project Managers often find that they "hear only what the project team wants them to hear," rather than the reality of the situation. For example, suppose that at the end of a given reporting period, tasks A, B, C, D, and E were planned to be completed. The planned costs to complete each task were determined to be A, \$10,000; B, \$35,000; C, \$47,000; D, \$28,000; and E \$22,000. However, only A, B, and C were completed; D and E were not started; and the actual cost to complete A, B, and C (or ACWP) turned out to be \$100,000. Note that the planned cost to complete A, B, C, D, and E, that is, the BCWS, is \$142,000, while the planned cost to do A, B, and C (or BCWP) is \$92,000. This calculation is easy to perform because A, B, and C are completed, in that some test was performed to ensure completion. Tasks D and E were not yet started. The difficulty arises in obtaining "realistic" cost estimates to determine the cost to complete project work still in progress. Let's consider, for example, that task C has not been completed. First, the Project Manager must determine how close the task is to completion by asking the assigned project personnel. Here is the dilemma. The new estimate may, indeed, be a guess on the part of those working on task C. How can the Project Manager really know whether the engineers or programmers working on the uncompleted task are telling the truth or just what they believe the Project Manager wants to hear? In order to minimize these problems and try to ensure accurate estimation of progress, full-scale performance testing should be performed on a monthly basis.

Variance Analysis. Figure 5.3-5 shows that the *Cost Variance* (CV) is equal to the difference between the BCWP and the ACWP:

$$CV = BCWP - ACWP \qquad (5.3\text{-}1)$$

The CV is the cost difference between the planned cost to complete tasks A, B, and C and the actual cost to complete the tasks. In the above equation, $CV = \$92,000 - \$100,000 = -\$8,000$, indicating that the project is \$8,000 over cost.

Note that in Figure 5.3-5 the *Schedule Variance* (SV) is denoted by *dollars* and not by *months*. The SV is defined as the difference between the BCWP and the BCWS:

$$SV = BCWP - BCWS \qquad (5.3\text{-}2)$$

The SV is given in terms of *dollars* because, if the project is behind, it will cost SV *dollars* to catch up. The SV is indeed an estimate of the amount of money needed. On the other hand, if an engineer were to report two months of slippage, the Project Manager still would not know how much money it would cost to catch up!

In the preceding example, $SV = \$92,000 - \$142,000 = -\$50,000$, it

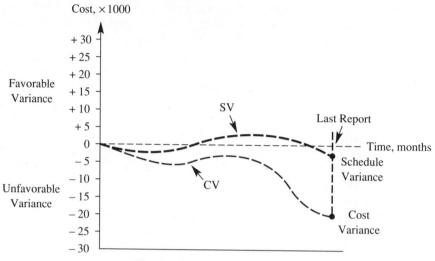

Figure 5.3-6 Variance analysis.

would take approximately $50,000 to catch up. Thus we are $50,000 behind schedule.

To illustrate the Variance Analysis, refer to Figure 5.3-6 and assume that at the last report the BCWP was $200,000. Referring to Figure 5.3-6, we see that the project is on schedule with peak deviations of less than $4000 or 2% of the BCWP. However, the CV is approximately $20,000. If not corrected, this high overrun could greatly impact the resulting product's profitability.

Each month, the Project Manager should receive the BCWS, BCWP, and ACWP for each task in level 2 of the WBS in the *Cost Schedule Status Report*. In Figure 5.3-7, level 2 consists of seven activities, tasks 1.1–1.7. (Note that the managers of each of these activities should receive a similar cost report that displays the level 3 activities.) For example, the manager of Task 1.5 receives a similar breakdown, including activities 1.5.1, 1.5.2, 1.5.3, and so on. The BCWS, BCWP, ACWP, SV, and CV are given, as are the planned BAC, the current *Estimated Cost At Completion* (ECAC), and the variance (difference) between them. In the next section, we explain how to obtain the ECAC and the *Estimated Time At Completion* (ETAC).

The Project Manager sees, on reviewing the Cost Schedule Status Report shown in Figure 5.3-7, that Systems Design is overspent and well behind schedule. The Project Manager also sees that this schedule slippage is beginning to affect the rest of the program. As a result, the Project Manager should speak to the Manager of the Systems Design activity to ascertain the difficulty and to help remove roadblocks that will enable task 1.2 to get back on schedule.

Performance Data

WBS Elements	BCWS	BCWP	ACWP	Variances		At Completion		
				Schedule (SV)	Cost (CV)	Budget (BAC)	Est (ECAC)	Var
1.1 Management	193,898	193,898	196,645	—0—	(2,747)	537,482	545,097	7,615
1.2 System Design	868,666	784,948	968,434	(83,718)	(203,486)	990,762	1,249,602	(256,840)
1.3 Testing & Evaluation	166,676	134,846	149,854	(31,830)	(15,008)	504,275	560,278	(56,124)
1.4 Training	7,909	7,909	2,225	—0—	5,684	124,475	35,018	84,457
1.5 SW Development	61,639	9,128	5,646	(52,511)	3,482	697,844	431,642	266,202
1.6 Support	7,908	7,908	2,224	—0—	5,684	23,882	6,716	17,166
1.7 HW Development	72,448	72,448	72,448	—0—	—0—	455,377	455,377	—0—
Total	1,379,144	1,211,085	1,417,476	(168,059)	(206,391)	3,334,097	3,281,851	52,246

Figure 5.3-7 Cost schedule status report.

Estimation of Cost and Time at Completion. The ECAC and ETAC are estimated monthly to help the Project Manager determine the impact of slippages.

The formula used to calculate the ECAC is

$$ECAC = \frac{ACWP}{BCWP} \times BAC \qquad (5.3\text{-}3)$$

Thus, the ECAC exceeds the BAC if we are overspending, that is, if the ACWP is greater than the BCWP.

Similarly, we find the ETAC to be given by the formula

$$ETAC = \frac{BCWS}{BCWP} \times TAC \qquad (5.3\text{-}4)$$

Here we see how long the project is expected to be delayed, when it is behind schedule; that is, when the BCWP is less than the BCWS.

5.3.3 Cost Estimation

The Project Manager is resonsible for preparing monthly cost reports such as the report shown in Figure 5.3-7, which is based on expenditures for direct project costs—labor, materials, and so on.

Typically, estimates are summarized by activities for specified time intervals and reported on or rolled up from the cost account level. To be of major value to project personnel, these reports must include:

- Activity and cost identifiers for each reported WBS work element.
- Activity descriptions.
- Actual activity start and finish dates and their durations for all completed activities.
- Actual start dates of activities in progress and their durations to date for activities in progress.
- Date on which progress is being reported.
- Amount paid for work performed on the activities.
- Calculated amount to be paid for the cost of the activities in progress.

Once obtained, cost estimate reports must be analyzed and interpreted in coordination with appropriate Functional Managers so that current and future deviations from the original budget plan can be controlled.

5.3.4 Cash Flow Forecasting

In today's capital-intensive project environments, it is essential to obtain a proper forecast of the project's cash flow.

Care must be taken to measure expenditures for direct labor, equipment, and materials and to record these payments throughout each project life-cycle phase. While the project budget serves as a tool for forecasting cash flow during the early phases of the life cycle, later on the Project Manager will need to put in place performance analysis techniques to monitor actual expenditures for completed project tasks. The Earned Value or Variance Analysis explained in Section 5.3.2 is an excellent control technique that lets Project Managers compare actual expenditures with historical or budgeted expenditures for individual activities or work packages.

Also, project cash flow reports, such as the status report shown in Figure 5.3-7, are schedule- or time-oriented and can be readily produced using most computer-driven *Critical Path Method* (CPM) software systems. The software systems calculate cumulative activity costs using the same programming algorithms that they employ to produce a CP scheduling network. By assigning costs to each of the activities associated with the network, direct labor costs over the duration of the activities can be summed and budgeted costs compared to the actual costs of completing project activities for different time periods.

Using standard programming techniques to determine the earliest and latest start schedules for each activity, cumulative total cost estimates and cash flow forecasts are projected.

Figure 5.3-8 is a graphic representation of total project costs. This report helps a Project Manager to predict cost overruns or underruns and control the associated constraints.

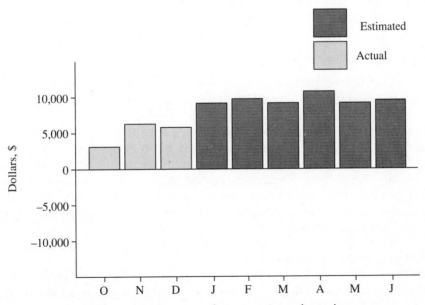

Figure 5.3-8 Project total costs, variance from plan.

5.4 TIME AND COST TRADE-OFF PROCEDURES

Schedule slippages inevitably overrun a budget. If a high-tech project is going into a market with a 20% growth rate, 12% price erosion, and 5-year product life, profits will be reduced by only 4% if development costs go 50% over plan. However, lifetime profits will be reduced by 33% if the project is 6 months late!

In Chapter 4 we explained how the Project Manager can use the CPM computational procedure to estimate total project costs for different completion dates. This procedure can also be used to control slippage and cost overruns.

The CPM computational procedure determines the critical path in the project and identifies the one that can be expedited with minimum cost. This activity is crashed and the process repeated until no savings will result from further shortening any activity.

As explained in Chapter 4, there are several assumptions regarding the relationships of activity costs and performance that the Project Manager should understand before attempting to employ this technique as a cost model. The following assumptions apply:

- The time–cost relationship is discrete.
- The time–cost trade-off points lie on a continuous linear or piecewise linear decreasing curve that can be crashed.
- Activities are independent; that is, the performance of one activity does not affect the performance of another.
- Buying time on one activity does not affect the cost of another.
- Resources are available for completing the project on a crash schedule.

5.4.1 Cost Optimization

By definition, the CPM procedure solves the project time–cost trade-off problem by reducing the duration of project activities at a minimum additional cost. To minimize activity costs, the basic calculation is altered to give a total cost for all activities as a function of time. Application of the same linear programming algorithm yields the familiar U-shaped plot of points illustrating the effect shortening activity length has on project costs.

Again, the Project Manager is cautioned. The same assumptions and limitations pertain to this application in that the method fails to allow for the shifts in the durations of noncritical activities that may increase the cost of work on the critical path.

How feasible is it, therefore, for Project Managers to employ this method as their only cost control technique?

Despite its immediate drawbacks, this method, when combined with other relevant monitoring and cost control techniques, can be a most useful decision-making tool.

5.5 ADHERING TO DESIGN SPECIFICATIONS—QUALITY

Quality is conformance to specifications. In the previous four sections of this chapter we explored techniques that can be used to monitor and control the schedule and cost performance of a project. Making sure that a project's technical performance complies with agreed-on requirements and standards (quality) is another of the Project Manager's major responsibilities. The assessment of how well the product in production adheres to its original design depends, in part, on how adept the Project Manager is at implementing a number of different measurement systems:

1. Performance testing and quality control procedures.
2. Frequent design review meetings.
3. Exception configuration and change control management.

5.5.1 Technical Reviews

In general, technical reviews are conducted by nonteam, project experts who examine the technical integrity and assess the quality of the product being built. Technical reviews often are classified for each of the different project life-cycle phases in regard to the basic purpose for which they are sponsored.

For example, reviews scheduled during the design phase are referred to as *design reviews* and are held primarily to aid the design team in soliciting recommendations from other technical experts to improve the product's design. Formal technical reviews, used to validate project objectives and assess compliance with design requirements, are known as *compliance reviews*.

Compliance reviews should be held at key technical decision points when authorized approval is needed to proceed to the next project phase. While design reviews answer the question of whether project deliverables and their associated milestones are valid, timely, and auditable, compliance reviews determine whether the product complies with the specified design requirements. Essentially, the purpose of this technical review is to analyze and recommend alternative solutions to the problems raised by the reviewers and the inspectors of working prototypes.

In MIS or software development projects, the technical reviews are often correlated with software inspections. Code walkthoughs are the inspection procedure by which the assembled project team reviews, as peers, all the different functions produced in the definition and design phases to determine what is good, bad, or missing. Minor problems, such as errors or typos in the code structure and other potentially costly mistakes, are identified and reported so that the software team can correct these errors early in the project design phase and reduce overall project rework.

To help assure quality execution of a software system's specifications, for instance, end users internal to the firm who have commissioned the work may audit, at key control points, the methods used to produce project deliverables. Internal auditors can also help to determine whether the system, as constructed, addresses at least minimum specifications by trying out working prototypes of the designed systems. Prototype demonstrations emphasize hands-on display of the working system rather than passive study of written documents. By properly conducting this form of user inspection, programmers can provide themselves with closer and longer user contact over the project life cycle, resulting in greater understanding of the permanent system.

5.5.2 Change Control Techniques

How does a Project Manager control changes in a project's technical design that affect the scheduled completion or delivery date of the product in development? How is the loss or gain of project personnel and other resources tracked and recorded? How are requests for new features and functions handled once project work has begun? Who is responsible for accepting formal requests for engineering design changes due to project rework?

In the best-managed projects, the Project Manager coordinates the activities of a formal Change Control Board vested with the responsibility for standardizing change procedures of all projects.

Change is defined as any alteration in baseline data that begins with a request and ends with a formal signing off on the approved documentation. For instance, if the design and testing of a specification not agreed to by the customer increases the cost of the project, the Project Manager must notify the customer and immediately draft an *Engineering Change Proposal* (ECP). The ECP documents the project's change in scope and allocates time and money for the modification. Engineering changes or design modification requests are usually classified as either "major" or "minor" (Class A or Class B) by the Change Control Board to clarify the imposed impact accepting or rejecting the change will have on the overall project.

It is necessary, therefore, that upper management be appraised of the requests in that any alterations to the product's design will most certainly either constrain resources assigned to project activities or delay the scheduled end date. The Project Manager must keep upper management informed of all baseline data changes and log each change as it is implemented into a change control ledger or computerized data base management system. Such systems can be formatted to produce change control reports and maintain historical data files concerning the different design modification requests and executed engineering changes. Historical data should be traceable to the work package level of the project's WBS so that the individuals initially responsible for the product's technical specifications can be kept informed of product enhancements or end-user requests for improvements.

Change is inevitable and should *not* be regarded as a problem in that it is controllable.

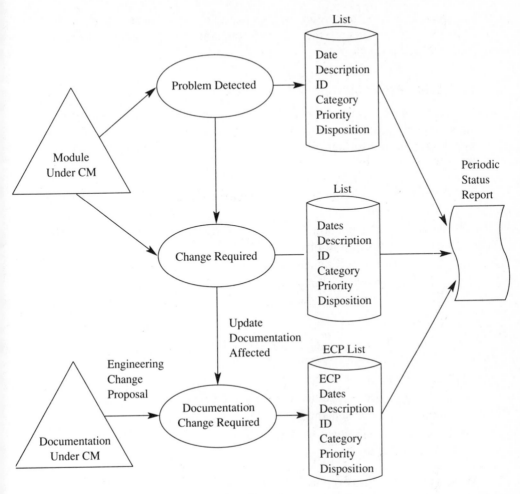

Figure 5.5-1 Configuration management.

5.5.3 Configuration Management

Closely related to the concept of engineering change control is the process of configuration management (CM). A system or product in development is said to be under configuration control when adjustments to the project's baseline schedule and budget reciprocate from accepted engineering design changes. As illustrated in Figure 5.5-1, configuration management is the discipline that integrates the adminsitrative activities of the project management process with the technical development work.

The discipline of configuration management ensures that the established baselines are reviewed whenever engineering or system changes are approved, and that the impact of the imposed changes on the overall project schedule, along with corresponding fluctuation in resource demands, be assessed, statused, and reported against the initial baselines. The new configuration is tracked thereafter throughout the remaining life-cycle phases of the project.

5.5.4 Status Reporting

The best report form is one that management will read and understand. Status reports should present control-oriented information in summary form and should be in a format requested by the sponsor. Among the important items to discuss are:

- Problem areas of concern to management and the probable cause(s) of any exceptions.
- Qualitative and quantitative impact of development statistics on the schedule, budget, and future profits.
- Action taken or recommended to alleviate the unfavorable impact.
- Indication of what management can do to help or to convey the exception to top management or notify the customer and other pertinent clientele.

The report might conclude with a set of exhibits that present:

- Budget or resource usage curves.
- Updated project schedule versus planned schedule.
- Summary master schedule.
- Technical development statistics, test results, working features versus expected, action items, open issues, etc.

An example of a status report is shown in Figure 5.5-2.

5.5.5 Project Review Meetings

Another widely successful technique to measure and compare progress with baseline data is the periodic project review meeting held by project officials. The purpose of a project review meeting is to:

- Assess common goals and objectives.
- Review progress and determine current status.
- Identify problems.
- Solicit recommendations for improvement.
- Follow up on previous action assignments.
- Communicate with fellow team members.
- Educate and/or motivate project personnel.

Monthly status report

Task Name ————————————

Task Leader ————————————

Reporting Period End Date ————————————

General Status and Summary of Activities (Including Subcontracting Activity):

Accomplishments (Deliverables, Milestones, etc.):

Potential Problems and Areas of Concern (i.e., Technical, Schedule, Cost Impacts):

Personnel Involved in Task:

Figure 5.5-2 Monthly status report.

To be effective, review meetings should be scheduled regularly at specific times when project personnel can attend. An appropriate agenda and advanced distribution of relevant data is mandatory. The Project Manager must clearly define the meeting's purpose and encourage participants' contributions to this regard. The team must work together to document any schedule or budget changes, report all technical findings, and publish a list of action assignments that specifies (1) what the recommended action is, (2) how problems will be solved, (3) who is responsible for their resolution, and (4) what the scheduled action will be and when it will be completed.

The project's technical success depends ultimately on the project team working together cooperatively to alleviate unavoidable customer delays, technical failures, and budget or schedule overruns due to unnegotiated changes in the scope of work.

5.5.6 Quality Control and Testing

Quality is conformance to specifications. The achievement of a project's technical objectives is assessed by using a variety of performance measures of which quality is foremost to success in the marketplace. Classical quality theory defines quality in terms of fitness for use. But as the leading consultants to industry who are recognized for their contributions to this field (Juran, Drucker, Demming, and Garvin) emphatically point out, fitness for use is determined by what the customers say it is—not upper management, marketing, or engineering.

Customer expectations ultimately shape a firm's quality policy. The initial steps taken during the definition and design phases of the life cycle by the project team to understand the quality characteristics desired by a customer also serve as the reference points against which the technical performance objectives are specified.

This commitment begins by planning quality designs and thoroughly testing their execution throughout product development.

Key to the design and development of quality products is management's focus on clear and articulate policies and standards. Corporate directives must be disseminated to Functional and Project Managers to ensure their adherence and implementation among the different business units contributing to project work.

These standards provide the framework for a product's functional requirements (specifications) and define how the project team should build, for example, its application programs and systems or communications hardware. As the customer's acceptance of product quality varies, management must make every effort to update its standards and policies regarding defects, reliability, performance, and maintainability (serviceability).

As shown in Figure 5.5-3, recognition of the importance of early and consistent control procedures to assure product quality is the first step

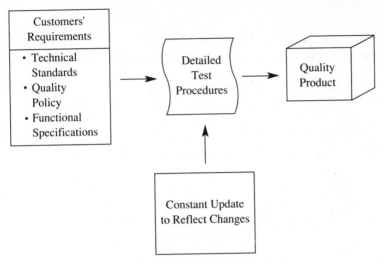

Figure 5.5-3 Quality assurance process.

toward carrying out a well conceived quality improvement process on a project. Project Managers must be willing to improve their team's performance by taking on the responsibility for preventing the perpetuation of design errors into development, prototype fabrication, and manufacture.

The following quality appraisal meaures help to assure the team's use of feedback on error and defect detection for design improvements of future product releases:

- Technical reviews and inspections.
- Regression and functional tests.
- Fault insertion tests.
- Performance and stress tests.
- Reliability and maintainability tests.
- Customer and end-user tests.

The Project Manager and the entire Project Team must cooperate fully to meet quality objectives. Each activity must be tested and compared to specifications. Only if all specifications are met is that activity complete. If this procedure is followed, then after system integration, the system test will yield only a few bugs, which can be traced and corrected. If the procedure is not followed, the system test will yield so many bugs that the debugging process will take an inordinately long time. "Do it right the first time" means correct your errors up front. Do not wait for system testing to correct errors.

5.6 DOCUMENTATION CONTROL

A timely and valid project control process includes provisions for the Project Manager to manage and control all the different documents associated with the project work.

Project communications must be controlled because the verbal transfer of information between developers is prone to the human failings of forgetfulness and misunderstanding.

The following lists many of the reasons why standardized control must be exerted over written and verbal communication in the project environment.

1. Checkpoints are required to measure the cost and technical performance of project work over time. The status information passed upward at regular intervals provides project leaders with a documented picture of how the team is progressing.

2. As project requirements change, the documents recording the agreed-on modifications also must be filed and maintained.

3. Initial and current product descriptions that detail every function and likely enhancement must be accessible throughout the development phases to every member of the design team.

4. Project tracking and reporting procedures along with corporate policy statements concerning the project review processes need to be communicated to all those contributing to project work.

5. Economic studies, technical specifications, and test plans, as well as the administrative and installation guides, provide project planners and designers with a common road-map against which to plan, integrate, and measure their work.

6. Technical documentation that accompanies the transfer of product from the development life-cycle phase into manufacture must be accurate and up to date as well as traceable to the engineers or software developers responsible for the design work.

Project documentation can be divided into two classifications or categories. One category is related to the actual technical work (specifications, implementation and design standards, test plans, etc.) The other documents project related business (work authorizations, reports on project status, signed agreements, jeopardy reports, and minutes of review meetings.)

Control of the documentation is achieved by using a Project Library or File that is found in the Project Office. The Project Library contains all documentation relative to the project along with previous schedules and other useful literature pertaining to how projects are implemented in the firm. This separate reference manual, often called a *Project Handbook,* describes how projects are conducted. It is also a useful source for specifying how to control and file the various technical documents that the developers generate.

It is somewhat difficult to distinguish between these two documentation categories in that the contents of some technical documents also pertain to project business. Certainly, one can classify project review meeting minutes as either technical or project-related documentation. The same is true of project reports and other signed agreements.

In general, the Project Manager need not worry about how project documents are classified but should, instead, concentrate on establishing either a manual or computerized data base mangament system to control and track what is filed.

5.7 CONCLUSION

Successful management of a technical project's ongoing performance is achieved by exercising daily control over the quality, schedule, and cost of work performed. Thoughout this Chapter, we have presented techniques for evaluating, measuring, and assuring proper implementation of a project's control system. In Chapter 6, we explore how the leadership and influence skills that a Project Manager has over the individuals who make the project happen ultimately contribute to its success or failure.

REFERENCES

1. Drucker, P. "Characteristics of Controls in Business Enterprise," *Management: Tasks, Responsibilities, Practices*. New York: Harper & Row, 1974, pp. 496–505.
2. Might, R. J., and Fischer, W. A., "The Role of Structural Factors in Determining Project Management Success," *IEEE Transactions on Engineering Management,* Vol. EM-32, No. 2, May 1985, pp. 71–77.
3. Might, R. J., "An Evaluation of the Effectiveness of Project Control Systems," *IEEE Transactions on Engineering Management,* Vol. EM-31, No. 3, August 1984, pp. 127–137.

BIBLIOGRAPHY

Arthur, L. J. *Measuring Programmer Productivity and Software Quality,* New York: Wiley, 1985.

Archibald, R. D. *Managing High-Technology Programs and Projects*. New York: Wiley, 1976.

Burril, C. W., and Ellswroth, L. W. *Modern Project Management: Foundations for Quality and Productivity*. Tenafly, NJ: Burrill-Ellsworth Associates, 1980.

Cleland, D. I., and King, W. R. *Systems Analysis and Project Management* (3rd ed.). New York: McGraw-Hill, 1983.

Davis, D. "New Projects: Beware of False Economies," *Harvard Business Review,* March–April 1985, pp. 95–101.

Delaney, W. A. "Management by Phases," *Society for Advancement of Management,* Winter 1984, pp. 59–65.

Harrison, F. L. *Advanced Project Management.* Hunts, U.K.: Cower Publishing Company Limited, 1981.

Kerzner, H., and Hostelley, D. "Budgeting for R&D Projects Part I," *Journal of Systems Management,* Vol. 35, No. 2, February 1984, pp. 6–11.

Kerzner, H. and Hostelley, D. "Budgeting for R&D Projects Part II," *Journal of Systems Management,* Vol. 35, No. 3, March 1984, pp 8–16.

Moder, J. J., Phillips, C. R., and Davis, E. W. *Project Management with CPM, PERT and Precedence Diagramming* (3rd ed.) New York: Van Nostrand Reinhold, 1983.

Sadek, K. E., Tomeski, A. E., and Kent, J. "A Late Project?: More Staff and Hardware Overutilization Makes it Later," *Data Management,* August 1981, pp. 14–19.

Swrette, G. J. "The AT&T Quality System," *AT&T Technical Journal,* Vol. 65, No. 2, March–Arpil 1986, pp. 21–29.

Wiest, J. D., and Levy, F. K. *A Management Guide to PERT/CPM: with GERT/PDM/DCPM and other Networks* (2nd ed.). Englewood Cliffs, NJ: Prentice Hall, 1977.

Case Study*

GEARING UP FOR A CRASH PROJECT

Introduction

For nearly a year, E&S Computing had rivaled the number one U.S. personal computer vendor to surpass its coveted position in the rapidly expanding personal computer market. According to Marketing, realizing this quest would mean releasing their new product by the third quarter of next year, in time for holiday sales and anticipated New Year business expansions. But delays in prototype development already were compromising E&S's likelihood of meeting its market window. Because the computing venture was considered so important to E&S's future, the company President requested that the development team get the project back on schedule whatever the cost!

Project Background

The laptop computer project had begun smoothly. In fact, the initial design work was not considered exploratory, and no delay was expected in accomplishing the specifications laid out by Marketing. It was originally believed that the product need not push any new technologies. Rather, the laptop was intended to be a downsized and more rugged personal computer than the larger desktop version introduced by E&S nearly a year ago. It was considered among the assigned engineers to be, from the start, a final development effort. Consequently, Product Engineering began work almost immediately to carry out Marketing's objectives and turn out a design that could be manufactured within 21 months. The intent was to beat the number one manufacturer to the market.

The project was composed of two groups of physical and electrical engineers who reported to separate organizations but who had worked together previously as a team. Because the project was small, it was felt that communications need not be bogged down by rigid reporting procedures. The project proceeded along rather informally until the first design review. By the time the chief engineers had gone through the specifications generated by the product planners, it became apparent that the original target dates and schedules had to be scrapped.

The laptop, it appeared, was to be all things to all consumers. In attempting to accomplish this project objective, Product Planning had failed to consult with Product Engineering. The design appeared so cut and dry that there was no large-scale implementation plan to steer the development

---

* This case study was written for class discussion purposes only. It is not intended to reflect practices or products of any particular company or organization.

Table I   Milestone Chart Laptop Computer Project

Task	Planned Duration[a]	Staff	Uncertainty[b]	Precedence
1. Electrical				
a. Electrical design	4	6	1	Start
b. Simulation	4	3	1/2	*a*
c. Construct breadboard	4	3	1	*b*
d. Testing	1	2	1/4	*c*
e. Final drawings	1	1	1/4	*d*
2. Application software				
f. Design	4	1	1	Start
g. Programming	2	2	1	*f*
h. System integration and testing	1	2	1/4	*g*
i. Documentation	1	1	1/4	*h*
3. Mechanical				
j. Chassis design	4	2	1	*a*
k. Construction	2	2	1	*j*
l. Final drawings	1	1	1/4	*k*
4. Assembly				
m. Complete computer	2	2	1/2	*d,h,k*
n. Quality assurance testing	1	2	1/2	*m*
o. Documentation	1	4	1/4	*n,e,i,l*
p. Turnover for manufacture	—	—	—	*o*

[a] Months.
[b] Standard deviation = months.

of this product. There existed only a list of initial milestone dates generated by Marketing (as shown in Table I), a Gantt chart of the schedule prepared by the planning engineers (Table II), and a personnel table that had failed to be updated since the project began (Table III). It was apparent that the product really lacked the technical approach required to engineer the design

Table II   Preliminary Grant Chart—Electrical Engineering Tasks

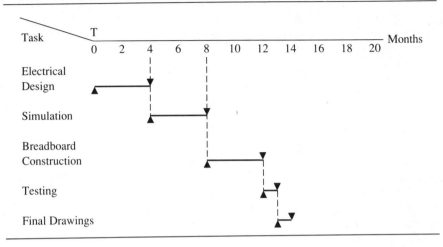

laid out by the planners. Somehow, the product's initial objectives had been misinterpreted. A more advanced gate array design would be essential to supply the features called for by Marketing in order to leapfrog the competition and place E&S into the number one seat. The planners had not been concerned with Engineering's input because they believed that the product was somewhat technologically uneventful. Senior engineers had not been assigned to the project, in anticipation of the ease of the initial design. Now the apparent rework and new design options generated many conflicts and delays the project could ill afford.

Table III   Staffing Profile—Electrical Engineering Tasks

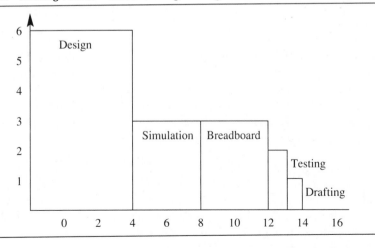

A Response To The President

A small task force of senior management decided, behind closed doors, that the laptop must be launched on time and that no expenses would be spared to accomplish this goal. The project's development budget would be increased and monies allocated to a single project organization, charged with the authority and responsibility to manage them. Future overruns would not be tolerated. This new project organization would need to exercise proper controls to manage the increased funds. All parties assigned project work—Engineering, Marketing, Planning, and Quality Assurance—would report to the new project group in the hope of facilitating better communication. To manage the work, the development team would report weekly to a project coordinator, Sue Johnson, who would tabulate all cost estimates and schedule projections. This time there would be no surprises. Management could direct its efforts to steering the product's new design instead of reacting to ill-conceived efforts that erode project team morale and misuse scarce corporate resources.

Organizing The Project

The second design review was scheduled for the first Tuesday in February, some 6 weeks after the new team began work on the portable computer. Sue had her job cut out for her, and she knew that proper controls were mandatory to manage LAP I—the code name conceived for the project by the new design team. This group would be responsible for the product's overall design and development as well as for monitoring the fabrication of the first tool-made samples in the factory.

Sue had requested that at least one senior hardware designer be assigned to LAP I to lead the less experienced electronic engineers through the project's initial life-cycle phases. Other key members of her team included a lead planning engineer and representatives from Manufacturing, Marketing, and Quality Assurance—who, she hoped, would open communications among the involved business and technical groups. Sue intended to formalize specific management procedures to plan, schedule, and control this project and immediately set about implementing her plan.

As Project Manager, Sue needed to report to management an estimate of how long and what it would cost to complete this project. To accomplish this objective, she called together the new LAP I team:

Ladies and gentlemen, welcome to the team! We're about to embark upon a new endeavor that will bring E&S to the forefront of the computing industry. In a little less than 2 weeks, top management will be expecting us to report on our design activities and predict whether we can get a product out and on the market within the next 10 months. All our market directives indicate that LAP I is what the business and

Table IV   Revised LAP I Schedule

Task	Duration After 13 Months	Staff	Uncertainty[a]	Precedence
1. Electrical				
a. Electrical Design	—	—	Completed	—
b. Simulation	—	—	Completed	—
c. Prototype	2	3	1	*b*
d. Evaluation & Testing	1	2	1/4	*c*
e. Final Design	1	1	1	*d*
f. 15 Boards Assembled & Tested	1	2	1/2	*e*
g. Drawings	2	1	1/4	*f*
2. Software				
h. Design	—	—	Completed	—
i. Programming	2	2	1	*h*
j. Unit Test & Evaluation	2	2	1/4	*i*
k. Final Test & System Integration	2	2	1	*j*
l. Documentation	1	1	1/4	*k*
3. Mechanical				
m. Design	—	—	Completed	—
n. Model made	—	—	Completed	—
o. Final design[a]	—	—	Completed	—
p. Final model	2	2	1	*o*
q. Drawings	1	1	1/4	*p*
r. Tool-made sample	2	2	1	*q,l,g*
s. Test & evaluation	1	2	1/4	*r*
t. Quality assurance	1	2	1/4	*s*
u. Documentation	1	4	1/4	*t*

[a] Months.

computing public is looking for in the next generation portable. It's apparent that we've a product to manufacture, so let's get going!

At this point in the project effort what's needed is a Master Schedule and Plan detailing every activity necessary for a successful product launch and the completion of this project. Therefore, I would like from each and every one of you a list of the work activities that you've assigned your staff and a report outlining the status of any outstanding assignments. We must begin immediately tracking the progress and completion of work assignments. In addition, if we're ever going to be able to measure accountability for project tasks, a detailed delineation of responsibilities is required. In other words, please consult your engineers and have them prepare a work breakdown structure (WBS) of their design activities for this project. The WBS should subdivide their work into manageable units or packages that we can review readily with top management.

In response, Sue's team went straight to work gathering the needed information to plan the project assignments and prepare the new schedule.

Table IV presents the activities, expected durations, and status of actually completed tasks collated by the LAP I team.

QUESTIONS

1. As a member of the LAP I project team, review the data shown in Table I. Prepare an early- and late-start schedule for all listed activities. Determine the staff months required to complete the work.

2. Taking into account the slippages revealed by your colleagues in Table IV, how many months are expected to complete this project?

3. Determine the probability of completing the project on time.

4. Using the revised schedule, prepare a three-level WBS; compare your results with those presented to E&S Management given in Table V.

The Tactics of Managing Change

The new project plans and revised estimates for completing the project work helped Sue and her team steer through the inevitable—a major design change to meet required market specifications.

It became apparent at the second Design Review that two options merited serious consideration. The first involved a review by LAP I designers of the computer's features in order to determine which ones, if any, could be eliminated or added in subsequent model releases. If the team were to adapt a modular design, the laptop could still enter the third-quarter market as a

Table V LAP I Work Breakdown Structure (WBS)

full-feature machine. Later, first-quarter releases, however, would incorporate the additional features such as software to manage memory and to speed disk-intensive applications on plug-in expansion boards and an optional electroluminescent flat screen. The team debated the feasibility and complexity of the proposed designs for the new portables and decided that Marketing's input must not be ignored.

The second option involved a technique known as "crashing the project," which required Sue and her team to calculate how many additional designers would be required to expedite the schedule so that the design of certain laptop features wouldn't be delayed in the laboratory.

The resulting technical specifications and a revised estimate of the budget needed to implement them were presented to top management shortly after the Design Review. The new designs brought the project team accolades and just as important, a pledge for the corresponding increase in funds to complete the work over the next 10 months. Now it became apparent that the LAP I team would need to implement a tracking system to monitor and report on their progress. So Sue and her team set about the task of installing a project management reporting system to monitor expenditures for project work through centralized data collection.

Each week an assigned member of the team collated the designers' most recent estimate of time remaining to complete project tasks. The revised estimates were then used to calculate any variances from the planned estimates and report progress to management.

QUESTIONS

5. Using the data given in Table I, plot the Budgeted Cost of Work Scheduled (BCWS) by the electrical engineers.

6. Using Table IV, plot the Actual Cost of Work Performed (ACWP) and the Budgeted Cost of Work Performed (BCWP) after 13 months have elapsed. What is the Cost Variance? The Schedule Variance? The Estimated Cost at Completion? The Estimated Time at Completion?

7. Now, measure the cost and schedule variances for the mechanical engineers after 13 months: plot the BCWS, ACWP, and BCWP. What is the Cost Variance? The Schedule Variance? The Estimated Cost at Completion? The Estimated Time at Completion?

8. Repeat these calculations for the software engineers.

The Importance of Proper Planning and Control

The LAP I team successfully overcame the variety of scheduling and planning difficulties imposed by Marketing's tight window. Key to their success was the firm leadership exhibited by Sue Johnson and her project coordinators, whom she carefully chose because of their years of design experience and excellence in an appropriate business or technical discipline. Her team constantly strove to maintain open communication among all involved parties and went about measuring and forcasting the results of any deviation from established benchmarks. When necessary, corrective action was taken to realize project goals. What was vital to the health of this project was the establishment of appropriate checkpoints against which to measure and report progress to those in a decision-making capacity who consequently acted on the information and replanned the work.

After several false starts and a near miss in sales projections, E&S had learned the value of properly planning and controlling its product developments. Eventually the LAP I went on to be a very successful business venture earning the company well over $200 million during the holiday season. LAP II and LAP III followed its introduction, as steady growth and strong demands for portable computers surpassed the analysts' early projections.

In the new year, E&S also won a government contract to supply the Defense Communication Agency with 100,000 portable personal computers—a bid secured because of the company's advanced designs and management prowess!

EXERCISES

1. Why is a measurement system necessary in a project control system?

2. Many engineers resist the application of any control system. List advantages and disadvantages of the use of project control.

3. Why are test plans so crucial to proper project control?

4. What different test plans can you think of?

5. What should a test plan include?

6. Explain why *timely* progress reviews are needed to properly control a project.

7. What is meant by the statement, "infrequent reporting can result in *unstable project control*"? Illustrate by means of an example.

8. Why is a Change Control Board so important?

9. A project has the precedence/cost relationship given below:

Tasks	Precedence Relationship	Duration (weeks)	Planned Cost to Complete ($)
A		4	$10K
B	A	6	12K
C	A,B	8	16K
D	B	4	8K
E	C,D	6	14K
F	D	8	16K
G	E,F	4	8K

a. Plot the precedence diagram and determine the critical path.
b. Plot the BCWS.

The project is monitored every two weeks.

	2	4	6	8	10	12	13	14
A	5K	5K						
	50%	100%						
B	0	0	4K	4K	5K	6K		
			25%	50%	75%	100%		
C	0	0	5K	5K	5K	5K		
			25%	50%	75%	100%		
D	0	0	—	—	—	—	8K	
							50%	
E	0	0	—	—	—	—	—	
F	0	0	—	—	—	—	—	
G	0	0	—	—	—	—	—	

c. The entries in the table shown above are the actual expenditures for each task obtained every two (2) weeks, and the percent of the task *completed*.

    1. Plot the ACWP and the BCWP on the same graph as the BCWS.

    2. How late is the project?

    3. Is the project over cost? By how much?

    4. What is the estimated date of completion?

    5. What is the estimated cost of completion?

10. If the BCWP > BCWS, is the project early or late? Explain.

11. If the ACWP > BCWP, is the project over or under cost? Explain.

12. If the ACWP < BCWS what conclusion, if any, can you draw? Explain.

CHAPTER 6

PROJECT LEADERSHIP

6.1 INTRODUCTION

A project's success is, in part, contingent on effectively managing the constraints of time, money, and performance specifications. A basic ingredient, determining *how* schedules will be met, *whether* costs will be accrued, and *what* level of quality will result, is the leadership and influence the Project Manager has over the people who make it all happen. The central figure in accomplishing project success *is* the Project Manager.

Project implementation typically relies on the integration of plans, schedules, and control mechanisms across all contributing functions. A Project Manager's success, therefore, is clearly contingent on effectively managing and influencing the diversity of personnel who are responsible for outlining and implementing project components. In fact, more than in any other form of organizational or managerial work, projects, because of their multidisciplined and highly interdependent nature, demand a leader with skills to motivate a project team and build a strong sense of commitment.

Leadership is more complicated in project organizations than it is in other forms. Project Managers are usually faced with the responsibility for project integration and implementation and the accountability for success as well as failure. Yet, because of their position within the structure of the organization, they may possess minimum "legal" or direct authority over project contributors. Project Managers often must cross departmental and divisional lines to attain the input and support necessary for project activities. Project leadership, therefore, involves not only understanding and working within the physical and political boundaries of a complicated organization but also entails a firm appreciation of the diverse needs of the professionals who are part of the project team. Determining the mechanisms by which to achieve the level of support or "buy-in" necessary for project success truly is *the challenge of project management*.

This chapter examines the importance of Project Manager leadership and influencing skills for effective project management and explores the techniques by which a Project Manager can develop support and commitment from the people needed to get the job done.

6.2 SKILLS OF THE PROJECT MANAGER

One cannot overemphasize the importance of the integration function in the Project Manager's daily work. Especially in a high-technology project effort, there is a pressing need to join the project's technical and organizational components into a total system. Unfortunately, however, the traditional, functionally organized company (or agency) grants authority largely on the basis of one's position within the "chain of command." Project Managers must work with and through Functional Managers and technical specialists, to ensure that project objectives are met. These horizontal and diagonal relationships offer the Project Manager little "traditional authority" to get the job done. The activities and functions necessary to manage a project, therefore, require a variety of interpersonal skills and informal sources of influence and power.

Often, promotion to a Project Manager's position is based not so much on one's excellence in managing people or even in projects, but on one's superior ability to meet technical demands. An excellent design engineer, a master of circuitry, or a top-notch developer frequently is given that first opportunity in management through the promotion to the position of Project Manager. Unfortunately, the skills and activities required to manage a project may be considerably more demanding than the more traditional manager's job. In many instances the newly appointed Project Manager must struggle against tough odds. Faced with little authority and a lack of managerial skills, new Project Managers find themselves in what may well be considered a "sink or swim" situation.

Just how can Project Managers make it through this overwhelming professional challenge? They must first recognize that how well project work is accomplished depends on their skills in fostering project integration and managing team performance. Table 6.2-1 summarizes the skills a Project Manager must possess to get results. As would be expected, the characteristics and skills needed for working with and through people predominate. Project Managers can greatly influence the work climate by their actions. They must possess a desire to get things accomplished with and through other people, rather than in spite of them. Concern for project team members, an ability to integrate the personal objectives and needs of project personnel into project goals, and the ability to create enthusiasm for the work itself create a project climate that is high in motivation, involvement, communication, and performance.

Another prime concern is that Project Managers understand that first,

Table 6.2-1   Project Manager Skills

- Technical expertise and general knowledge

- Forecasting, planning, scheduling, and estimating

- Problem identification and solution

- Ability to establish objectives, performance criteria, and standards

- Big picture orientation

- Task organization

- Flexibility and adaptability

- Team selection

- Team building, management, and leadership

- Accounting, budgeting, and financial control

- Training, development, and delegation

- Communications: interpersonal, written, and oral

- Conflict resolution and negotiation

- Group dynamics and organizational development

- Creativity and conceptual overview

- Systems and project management

- Walk on water

last, and always they are *manager* of the project. There is a heavy temptation for Project Managers to practice their technical discipline (e.g., Hardware Engineering, Software Systems Analysis) throughout the project rather than manage the *process* of the project itself. This is a classic reason for project failure. In this case, not only are project planning and team building functions sorely neglected but the project also suffers from a "myopic" or narrow technical view.

A Project Manager's skills should be varied. At best, the Project Manager should be considered a technical "generalist" who can relate to the greatest number of team members and attend to the broad requirements of the project. With the exception of very small projects, where the Project Manager may be required to participate both as a technical contributor and as the manager of the project, the Project Manager must demonstrate and practice an understanding of human behavior and deal with not only technical but also highly charged *emotional* issues.

6.3 AUTHORITY AND INFLUENCE—THE KEYS TO EFFECTIVE PROJECT LEADERSHIP

Working with and through others to accomplish organizational objectives is the cornerstone of *any* management position. A major difference, however, between the more contemporary Project Manager's role and the traditional Functional Manager's position is the authority and influence base each possesses or fails to possess.

Project Managers often operate in a complex, multidisciplined environment with little control over contributing functional specialists. Lacking the traditional "boss–subordinate" relationship, Project Managers must frequently rely on developing the *perception* of authority as well as influence strategies that are derived from a variety of interpersonal and political sources. The first step in influencing others is for Project Managers to become aware of the sources of power and influence that are available to them as they operate across the multiproject, multidisciplined, and multichaotic organization.

To deal effectively with the multiple chains of command as well as the conflicting priorities that are characteristic of most project matrix organizations, Project Managers must be able to communicate many things: specifically, that they respect the concerns and perspectives of others and welcome and encourage views different from their own and, above all, that they can be trusted. Because they lack direct control or authority over the people performing the work, Project Managers must give and request support by using not only logic but also a variety of interpersonal influence bases.

6.4 AUTHORITY, RESPONSIBILITY, AND ACCOUNTABILITY— THE PROJECT MANAGER'S DILEMMA

Our traditional concepts of authority, and the power it generates, stem from the following "formal" definitions:

> *Authority.* Traditionally, this term refers to the organizationally derived right to make the necessary decisions that others are *required* to follow. This includes issuing orders, determining directives, outlining missions, setting priorities, and scheduling target dates.

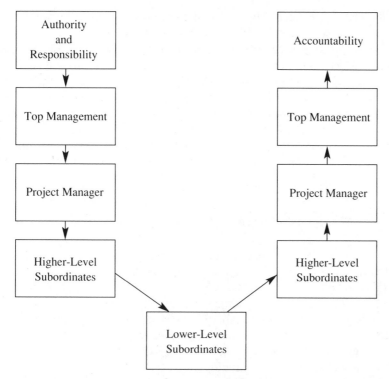

Figure 6.4-1 The ideal flow of authority, responsibility, and accountability.

Responsibility. We generally think of "responsibility" as consisting of the tasks, obligations, or activities which result from the specific assignments that are delegated in accordance with formal organizational roles or positions.

Accountability. Once our integrity or professionalism is tied to the successful or unsuccessful completion of a specific task or assignment, we can consider ourselves "accountable" for the job. Most managers for example are "accountable" for an employee's decisions, even though the task and the authority necessary to accomplish that task has been fully delegated.

Generally, "responsibilities" are derived from our job descriptions and, therefore, from one's formal role within the organization. "Accountability," although also somewhat organizationally derived, relies more heavily on one's sense of professional integrity and liability. As Figure 6.4-1 indicates, both authority and responsibility may be conceived of as flowing down the hiearchy or chain of command, as they are delegated from one level of the organization to the next. "Accountability," on the other hand, or the notion of "the buck stops here," moves up the organization, usually landing in the lap of one with a formal role or authority position, who can make the necessary changes to improve design or clarify objectives. In the case of project management environments, accountability for project success or failure generally flows *horizontally* as well as *diagonally* across the organization until it reaches the Project Manager. The traditional dilemma of the Project Manager, therefore, is to be able to work effectively within an organization that grants this important position *total responsibility, total accountability,* and yet *minimum authority.* The Project Manager truly must learn to rely on a variety of informal sources of authority to manage a project effectively.

6.5 TRUST AND INFORMAL CONTACTS

It is common for project team members to relay only what they believe the Project Manager wants to hear. Many organizations have reinforced the fear of sharing problems by often "killing the bearer of bad news." Project team members will share their problems with the Project Manager only if a sense of trust has truly been established; that is, if they truly believe that the Project Manager understands their problems and the difficulties they face, such as meeting what appears to be an impossible deadline or attempting to implement state-of-the art technology.

To be truly effective, a Project Manager must know the people involved, who they are, and what their problems might be. Project Managers must develop and use techniques to dig beyond such surface indicators as reports and forms, schedules, and budgets. They must find out just how the team is working together, what their technical as well as organizational problems are, and what issues or conflicts are preventing the realization of project objectives.

Getting to the heart of the issues and talking about what is really going on allows project team members to feel that a Project Manager not only has knowledge of the "big picture" but also has taken the time to become sufficiently familiar with team problems to be able to solve them. Such actions demonstrate a caring position and foster the development of trust in relationships.

6.6 DEVELOPING PROJECT AUTHORITY AND INFLUENCE

A variety of studies have focused on identifying several influence bases used by Project Managers in generating support from functional team members.

Hodgettes [1], in an investigation of a sample of firms in aerospace, construction, chemicals, and state government found that competence, persuasive ability, negotiation skills, and reciprocal favors were noted as important in overcoming what has frequently been referred to as the Project Manager's "authority gap."

Subsequent investigations by Gemmill, Wilemon, and Thamhaim [2–4] found that eight influence bases were available to Project Managers, contingent on the specific organizational form and project charter. These influence methods, rank-ordered (1 = most important, 8 = least important) according to support personnel's perception of effectiveness, include:

Influence Source	Mean
Authority—the legitimate hierarchical right to issue orders	3.0
Work challenge—the Project Manager's ability to capitalize on a worker's enjoyment of doing a particular task	3.2
Expertise—special knowledge the Project Manager possesses and others deem important	3.3
Future work assignments—the Project Manager's perceived ability to influence a worker's future task assignments	4.6
Salary—the Project Manager's perceived ability to influence or increase monetary remuneration	4.6
Promotion—the Project Manager's perceived ability to improve a worker's position	4.8
Friendship—friendly personal relationships between the Project Manager and others	6.2
Coercion—the Project Manager's perceived ability to dispense or cause punishment	7.8

As may be expected, each of these eight sources of influence has a different effect on the morale and climate of the project team. In fact, it was discovered that the *less* Project Managers emphasized organizationally derived influence bases, such as authority, salary and coercion, and the *more* they relied on work challenge and expertise, the higher they were rated by project support personnel in their ability to effectively manage projects. In fact, the use of position authority (as in muscling or strong-arming techniques) by Project Managers as a means of influencing support personnel led to lower ratings with regard to overall project performance. Open communication and task involvement among project participants, on the other hand, were positively associated with higher project performance [5].

While position authority may be regarded as an important basis of influence, it must be used judiciously and in accordance with the demands and characteristics of the particular project situation. Although perceived to be an extremely important influence base by Project and Functional Managers alike, research indicates that Project Managers who are perceived to rely a great deal on their position or legal authority to accomplish project objectives are rated by peers as well as followers to be *less* effective in their ability to resolve project problems and conflicts and are rated lower in overall project performance [5].

These investigations do not deny the importance of delegating as much organizationally derived authority to the Project Manager as the environment permits; however, once legal authority is available, a Project Manager will best be served by developing expertise, interpersonal skills, and work challenge. Moreover, although *technical expertise* is still widely recognized as a potential basis of influence and, as shall be discussed, is an important source of power, this expertise can be an effective source of influence only if it is *recognized* and *actively sought after* by both project team members and colleagues. Technical expertise, therefore, can serve as a potential influence base only if it is associated with effective lateral communication and the development of a wide spectrum of organizational contacts and interpersonal relationships. Thus, especially in the matrix environment, *technical expertise is effective only when associated with well-developed interpersonal skills*.

6.7 UNDERSTANDING THE BASIS OF PROJECT LEADERSHIP

The authors hope that at this point in our discussion it has become increasingly evident that to be effective in a multiproject environment, Project Managers and team players must posses skills that go beyond technical analysis, and cost and schedule estimating. Leadership abilities and interpersonal skills are critical to effective project performance. Project Managers, especially those who find themselves in a matrix environment, are in a very challenging leadership position. The better they understand the nature of leadership and how it relates to project efforts, the better they will be able to accomplish project objectives.

Effective leadership occurs when a "symbiotic," or a mutually growth-producing, relationship is created between a leader and followers. Leaders attempt to achieve or accomplish certain organizational goals, or even personal objectives, through the activities of their followers. The followers, in turn, actually grant or bestow the right to lead to an individual who offers them the fulfillment of their own objectives, which may include such things as professional growth, job challenge, respect, or a sense of integrity. The specific leadership style practiced by the Project Manager may, indeed, vary according to the demands of the task or the timing of the situation, but must *always consider the needs of the followers and provide opportunities for follower growth and development.*

The concept that leaders are born—not made—is a thing of the past. The "trait theory" of leadership, which maintained that certain characteristics such as height, eye color, or intelligence were necessary determinants of leadership, has been disproved time and time again. No consistent pool of traits or innate characteristics are available that indicate success or failure in a leadership role. Behavioral scientists have found that leadership is, at times, a phenomenon that is actually bestowed on or granted to an individual by the group itself. Take any newly formed task group of four or five people who have never met or worked together before, and *still* a leader evolves. We can observe and actually measure leadership behavior, for example, by the number of outputs or initiatives an individual makes (or better still, is *permitted* to make) toward accomplishing the group's task. Leadership, therefore, may be perceived of as a function of the number of outputs an individual makes that are actually incorporated into the group's decision(s). Thus, leadership is actually the ability to influence others to accept a particular idea, concept, or procedure that the leader holds true.

6.8 DEVELOPING A PROJECT LEADERSHIP "STYLE"

As we have previously mentioned, the authors consider project management to be one of the most participative management strategies in today's corporate environment. To be truly successful, projects must be integrated across corporate divisional lines through group consensus-seeking activities that produce a sense of "buy-in" and team commitment. This is accomplished through the contribution of all project-related disciplines to project planning, scheduling, and controlling processes. However, let us not confuse a participative management *philosophy* with a participative *leadership style.*

Project management techniques should be built on a general foundation of cross-functional, organizational participation. Specific project leadership *style,* however, is contingent on the combination of a number of complex elements. To be *participative* (or democratic) at *all* times certainly would be ridiculous! In case of a real crisis or safety hazard, for example, only weak and ineffective Project Managers would turn to their peers or followers for a vote. Project Managers must develop a *variety* of leadership styles that permit them to appropriately respond to the changing and complex challenges of a dynamic life cycle. In other words, Project Managers need to recognize that no one leadership style will work for all project conditions or situations. Although difficult, Project Managers must be able to diagnose a variety of situations and adapt their behavior or leadership style to fit not only the changing life-cycle demands of the project but also the requirements and style of the client, as well as the professional needs of project team members.

The concept of developing effective leadership through varying styles and behavior is not a particularly new one. Back in the early 1950s, Fiedler [6], in his "contingency approach to leadership," recognized that no one best leadership style existed and that different situations call for different approaches. Fiedler found that three important variables had tremendous affect upon leader performance:

1. *Position Power.* This refers to the degree to which the *power of the position* enables leaders to secure compliance with their directives. This is power arising directly from organizational authority. Fiedler points out that the leader with clear position power can more easily obtain followership than one who does not have such power.
2. *Task Structure.* The clearer the tasks to be performed, the greater the leader's power and control, and the easier (or more favorable) the leadership task.
3. *Leader–Member Relations.* Fiedler regards this dimension as the most important from the leader's point of view, since position power and task structure may be totally, organizationally determined. This dimension concerns the extent to which members trust, respect, and like the leader and are willing to follow.

The condition of each of these three variables determines how favorable (or unfavorable) a situation is for the leader. A most favorable situation is one in which the leader is well liked and respected by followers, is granted position power, and directs a clearly defined and well-structured task. On the other hand, a somewhat unfavorable situation is one in which the leader is new to the followers and, therefore, has not established clear-cut positive relations with followers; or perhaps is known to them but not very well liked or respected, has little position power, and is faced with the responsibility of directing a rather complex, unstructured task with fuzzy objectives or goals—a situation not far removed from state-of-the-art engineering projects!

Fiedler concluded that when each of these conditions is either *extremely favorable* or extremely *unfavorable,* a leader can be successful by demonstrating a highly task-oriented or directive style. When there is a mixture of conditions, however, such as the situation one frequently encounters in a dynamic project environment, people who are oriented toward relationships tend to be the more successful leaders!

6.8.1 The Situational Leadership Model

Hershey and Blanchard [7] more recently expanded on Fiedler's basic "contingency approach to leadership" and developed what they refer to as "Situational Leadership" theory. This theory proves to be extremely useful and applicable to the dynamic and changing requirements of a project effort. The researchers not only address the importance of the leader's own style (be it task- or people-oriented) in accomplishing group goals but also recognize that the needs, goals, and task experience levels of the followers must be appropriately addressed if the leader is to experience any success in the position.

Hershey and Blanchard's Situational Leadership theory considers the interaction of three major variables that determine leadership style and its appropriateness. These three variables are:

1. The *task behavior* or amount of direction a leader provides in the way of telling people what to do, where to do it, and how it should be done. This behavior is characteristic of leaders who set goals for their followers and actually outline the necessary steps to accomplish these goals.
2. The *relationship behavior* or the degree of "stroking," feedback, and support a leader demonstrates, often characterized by two-way communication and active listening.
3. The *maturity level* of the followers with regard to their knowledge or experience in relation to the specific task at hand. Maturity levels are dynamic in nature and can be determined in part by followers' needs for independence, readiness to assume responsibility for decision-making, their tolerance for ambiguity, and their need for feedback.

The interaction of task behavior, relationship behavior, and maturity level of followers, therefore, determines the appropriate leadership style necessary for accomplishing project objectives in a particular life-cycle phase.

Situational Leadership theory supports the authors' premise that effective leadership is a symbiotic or *mutually gainful* relationship between follower and leader; that is, to effectively accomplish project tasks, Project Managers must not only be aware of the characteristics of the task and the needs and goals of their followers but must also vary their leadership style to provide opportunities for follower growth, development, and "maturation." Figure 6.8-1 portrays the relationship between follower maturity and leadership style.

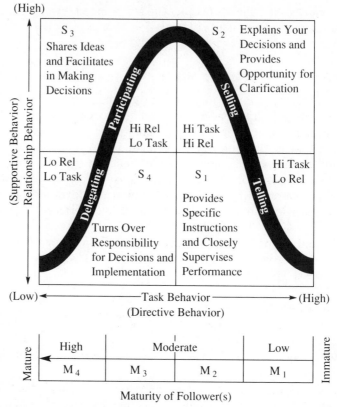

Figure 6.8-1 Determining an appropriate leadership style. From Paul Hershey, Kenneth Blanchard, and Walker Natemayer, *Situational Leadership, Perception and the Impact of Power,* The Center for Leadership Studies, 1979.

As indicated in Figure 6.8-1, four basic leadership styles occur within the context of three major variables: leader task orientation, leader relationship orientation, and the maturity level of followers. The four basic leadership styles and their appropriate behaviors are presented in Table 6.8-1. They are:

Telling (S_1). This style of leadership is characteristic of a leader who provides a great deal of guidance, direction, and input into the *decision-making* process regarding *what* tasks are to be accomplished, *how* the task shall be accomplished, and *where* the work is to be carried out. *Directing* or *guiding* leaders define roles and responsibilities for their followers and provide clear and specific instructions regarding technical methodology and procedures. This style is usually quite appropriate when faced with followers who may be unable to take

Table 6.8.1   Situational Leadership Styles[a]

S_1	S_2	S_3	S_4
Telling	Selling	Participating	Delegating
Guiding	Explaining	Encouraging	Observing
Directing	Clarifying	Collaborating	Monitoring
Establishing	Persuading	Committing	Fulfilling

[a] From Hersey, P. *The Situational Leader,* Center for Leadership Studies, 1984.

responsibility for task activities because of, perhaps, limited knowledge or experience.

Selling (S_2). Also referred to as the *exploring, clarifying, or persuading* style, the *selling* style of leadership is so called because, although direction, information, and knowledge is still provided mostly by the leader, a greater attempt is made to explain why certain actions need to be taken or methodologies followed. The leader further attempts to get followers to accept or "buy into" the decision by pointing out how their suggestions have been addressed and why this decision may be most beneficial to all. This style is not only appropriate for developing followers with relatively limited knowledge yet growing enthusiasm but also may be successful in selling concepts, proposals, and ideas to top management, as well as to the customer.

Participating (S_3). As the knowledge, experience, and overall organizational maturity levels reach a comfortable and relatively moderate range, the effective leader utilizes the group's ideas and expertise to formulate strategies and arrive at decisions. This style is frequently referred to as *participating, encouraging, collaborating,* or *committing* [8] because the leader and followers share in the problem-solving, decision-making process, with the main role of the leader being facilitator.

Delegating (S_4). As followers reach optimum maturity levels they experience both the motivation and ability to accomplish project tasks with minimum supervision or support from the leader. In the advanced stages of project design, for example, participants may be so familiar with specifications and methodology that the leader or Project Manager may assume a *monitoring* or *observing* role, providing the necessary opportunities for the project team members to fulfill both their task and personal objectives.

Each leadership style, therefore, can be regarded as having its "appropriate" place in terms of the life cycle of the project. As the level of follower maturity continues to increase over the life cycle of the project, leaders must reduce their task directive behavior by delegating more decision-making authority to project specialists. As the bell-shaped curve in Figure 6.8-1 indicates, effective leaders vary their leadership style along the curvilinear function as followers develop their project-related task maturity [8].

Project and Functional Managers alike must bear in mind that to successfully manage a project they will lead a variety of followers, including project team specialists, the customer and top management. The concept of maturity refers not merely to overall experience, educational level, or the need for feedback and autonomy, but to the level of understanding regarding specific technical, administrative, or monetary requirements. Customers, for example, may well believe that they fully understand the requirements of the system they wish to build, but may need to be "sold" on realistic cost requirements. The CEOs of a leading R&D firm may well have advanced technical degrees and broad-based administrative experience, but may need "clarification" regarding the success and feasibility of new project technologies.

The key to effective project leadership, therefore, is to assess the maturity needs of followers, whoever they may be, and behave as the model prescribes. Project and Functional Managers involved in the accomplishment of a state-of-the-art project, for example, will find that their engineers and technical specialists will vary greatly in maturity not only over the duration of the project, but in terms of the specific requirements of the particular task at hand. An experienced design engineer may master the technology well enough to require minimum input and direction, for example, when determining specifications or estimating resource requirements, but may not have demonstrated the same degree of "maturity" in documenting specifications or reporting progress to management. Thus, it may be quite appropriate for a Project Manager to offer little technical direction during the design phase yet provide a great deal of supervision over report writing activities or the process of reporting to management.

6.9 POWER—THE LEADERSHIP POTENTIAL

If we agree that leadership is "the ability to influence others towards the accomplishment of organizational goals or objectives," then power may be seen as "a leader's influence potential." It is the means by which influence is accomplished.

Indeed, it is difficult to separate the concepts of leadership and power. It would be impossible to influence another without utilizing some degree of power. Power is a feature of most organizational interactions and, at some time or another, is used by all employees to control scarce resources, negotiate agreements, and establish and/or reach professional as well as organizational goals.

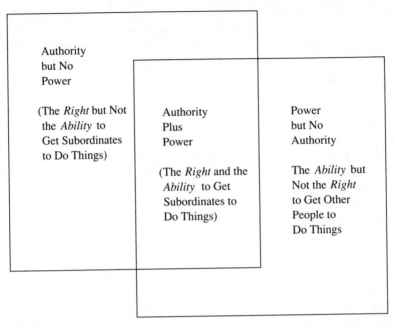

Authority
but No
Power

(The *Right* but Not
the *Ability* to
Get Subordinates
to Do Things)

Authority
Plus
Power

(The *Right* and the
Ability to Get
Subordinates to
Do Things)

Power
but No
Authority

The *Ability* but
Not the *Right*
to Get Other
People to
Do Things

Figure 6.9-1 The relationship between power and authority. From Robert Kreitner, *Management,* Houghton Mifflin, Boston, 1983.

Individuals may possess substantial power although they hold no formal leadership position. As previously defined, legal authority is conceived of as "the *right* to make final decisions that others are *required* to follow." Authority, therefore, is a formally sanctioned privilege that may or may not be available to all employees. Power, on the other hand, implies an *ability* to get results. As illustrated in Figure 6.9-1, one may conceivably possess authority yet demonstrate no power, possess no authority and demonstrate power, or, ideally, possess both authority and power.

6.10 IDENTIFYING SOURCES OF POWER

Of particular interest to Project Managers, especially those working in a matrix environment, are the various sources of power that may help to increase their influence both within the project team and with the customer. The more that Project Managers understand power and are better able to locate situations in which it arises, the more effective they will be in controlling resources and establishing their leadership position with the project team.

Hershey and Blanchard [7] expand on French and Raven's [9] five traditional bases of power to include the following:

Coercive Power. This power base is derived primarily from the use of fear tactics and is associated with the threats of punishment by, for example, an undesirable assignment, demotion, or a poor performance review. A Project Manager, for example, may warn team members that their performance review may be negatively impacted if the specification documentation is late.

Extensive use of coercive power may create a situation in which employees may, indeed, meet deadlines or document designs. However, such threats generally tend to inhibit risk-taking and creativity, which is the basis of innovative behavior. Continual use of coercive power may eventually lead to a breakdown in whatever influence and authority a Project Manager possesses. Strange and unique "illnesses" which threaten the very survival of the project have been known to inflict project teams working under coercive Project Managers!

Connection Power. Building alliances with influential or important people within the organization is an important power base for any manager, but especially for Project Managers who must work with and through functional personnel to achieve project objectives. Developing a variety of *informal contacts* can help Project Managers to be in a better position to recognize project pitfalls early. Technical people talk about their technical problems with technical colleagues whom they trust (i.e., with whom they lunch, play tennis, or have dinner). Project Managers should create a variety of informal contacts via breakfast, lunch, and dinner appointments, sports activities, and by conducting visits outside project areas, such as the manufacturing or subcontractor's facilities.

Expert Power. Knowledge and experience continues to rank high among influence and power bases, especially in a high-technology R&D environment. If project managers have prior experience working in contributing technologies, functional specialists may perceive them as competent and respect them because of the extra knowledge and experience they may bring to project efforts. Expertise, however, is not limited to technical knowledge alone. Effective Project Managers, furthermore, are seen as having the ability to solve *both* technical and organizational problems. Power comes from an expert understanding of adminsitrative methodologies and controls, as well as a firm grasp of customer needs. While it may be a difficult if not impossible task, finding a Project Manager with technical expertise, a broad knowledge of the organization, and an understanding of customer needs and requirements contributes greatly to project success!

Legitimate Power. Legitimate power is derived from an individual's formal position or location within the organizational hierarchy. The higher the position, the greater the power or influence associated with it. This power evolves from employee expectations and attitudes regarding certain roles within the corporation or agency. Legitimate power, therefore, simply is the "right to command." It may vary in its degree of potential influence depending on the values and expectations of the work group as, for example, when we compare a military agency to a private corporate work team. In recent years, we have witnessed an erosion of the effectiveness of legitimate power, perhaps through a combination of its frequent misuse, changes in societal values or demographics, and the increased educational levels of our work force. Formal positions held within the organization no longer are sufficient bases of influence. Project Managers operating within a matrix structure, moreover, tend to lack such formal position authority and would be wise to develop some of the other more informal bases of power discussed.

Referent Power. "Charisma" or personal characteristics has been known to influence the behavior of others simply because of the desire to identify with the leader. This power base is strongly associated with feelings of respect, admiration, or liking, which may result from a leader's past accomplishments or simply be a function of magnetism or personality. We have frequently heard the comment that charisma, like leadership, is something that simply can not be developed. Our stance on the development of leadership and all its related skills by now should be quite clear. Leadership is, like any skill, an ability that not only can be developed, but needs continuous fine tuning. Charisma, likewise, may be thought of as a function of effective interpersonal communication and listening skills. Developing such abilities assists the Project Manager in building *referent power.*

Reward Power. One basic source of power a leader may use is the ability to provide something of value to those who fulfill the project's objectives. For a reward to be effective, however, it must be properly aligned with the values of the person or people upon whom it is bestowed. In other words, a Project Manager must be familiar with the needs and values of the functional specialists working together to accomplish project tasks. Since monetary rewards may not always be available, the Project Manager must consider a variety of rewards that have the potential of satisfying diverse employee needs. Research suggests that while traditional rewards such as salary and promotion are important to technical professionals, of equal importance is *work challenge* and *recognition* [10]. The ability to provide opportunities for project team members to achieve these rewards will create a significant power base for the matrix Project Manager. Project Managers must go to great lengths to ensure that the work of team members is recognized and communicated through the formal and informal influence networks. Such actions and rewards may include a letter of appreciation or recognition, a recommendation for a promotion, or a merit increase. An important point for a Project Manager to remember is that empty promises of forthcoming rewards serve only to demotivate project team participants and erode trust. Project Managers must be perceived to have the ability to grant rewards, be they monetary or otherwise. Reward power, therefore, is clearly associated with *both* position and connection power.

The concept of rewards is, additionally, linked to the effective blending of a project team. Team rewards should be provided in celebration of the accomplishment of hard milestones and successful project team performance. Rewards help to reinforce the concept of *unity of mission* and provide an important sense of accountability as a functioning unit. Unfortunately, most organizations today advocate team performance yet reward their employees on the basis of individual contribution to unit goals (i.e., "What did you do this year to help achieve the unit's bottom line?" rather than "How did you contribute to project team performance?"). Failure to review performance on the basis of team contribution undermines the accountability of a team and contributes to poor project performance. Many of today's corporations have begun to recognize this important point and have begun encouraging peer reviews as a means of emphasizing the interdependence of team players and the tremendous importance of team accountability.

Information Power. Information concerning the project, its related market, or its relationship to other corporate or agency priorities that is perceived to be valuable to other project participants is an excellent source of influence. Unlike other bases of power, however, this source truly can only be effective if *shared* with others. Many individuals erroneously believe that their power base may be strengthened by hoarding information. This is *not* the case! Project Managers who hoard information will only serve to *discourage* individuals from seeking them out and therefore will eventually erode their power base. Information works to influence others because people want to be kept informed of, or in on, project happenings and made to feel as if something important is being shared with them. The Project Manager would be wise to develop many sources of information concerning the project as well as its internal and external environment. Verbal and written sources of information, such as speeches, reports, correspondence, journal articles, and newsletters place the Project Manager in a *central information processing position.*

Table 6.10-1 provides an overview and description of the various power bases available to Project Managers.

Table 6.10-1   Project Manager Power Bases

Power Base	Source	Comments
Coercion	Fear and the avoidance of punishment and threats	Use of coercive power is linked to organizational position; tends to inhibit creativity and negatively affect project team morale
Connection	Alliances with influential or important people	Highly effective for developing trust and recognizing project pitfalls early
Expert	Knowledge and experience, especially in a contributing functional area	To be effective, must be perceived by project participants to be vital to project success; limited if not expanded to the adminsitrative and customer knowledge arena
Legitimate	Formal position "right to command"	May vary in degree of potential influence, depending on the values of the group; less effective due to frequent misuse, changes in societal values, and increased education of work force
Referent	"Charisma" or personal characteristics	Highly effective; may be conceived as a function of interpersonal communication, and listening skills
Reward	Ability to provide positive sanctions for performance	To be effective, must properly correspond to participant values and/or expectations; since money is not always available, Project Managers must consider a variety of potentially satisfying sanctions especially those related to work challenge and recognition.

Table 6.10-1 *(Continued)*

Power Base	Source	Comments
Information	Knowledge or "tidbits" concerning project activities and related occurences	Effective only if perceived valuable and shared appropriately with functional managers and project participants; erodes trust and creates resentment if hoarded

6.11 DEVELOPING PROJECT MANAGERS

The Project Manager as the focal point of the project has bottom-line responsibility for making it all happen. They must be superb planners, super salespeople, artful negotiators, and talented psychologists! Especially in an high-technology R&D environment, Project Managers' perceived power within the organization and with the project team is highly related to their technical know-how and their skills of persuasion and negotiation. In the extremely demanding project environment, Project Managers who rely primarily on their formal authority, formal rewards, or whatever coercive power they may possess will be perceived as less effective than those who attempt to influence the course of events with their ability to deal with each situation as it occurs and who increase the probability of project success through integrative planning and organizational interfacing.

Project Managers must find ways to enhance their authority through establishing contacts and relationships both inside and outside the organization. Relationships must be established with such entities as staff groups, functional departments, superiors, the client, subcontractors, consultants, and, at times, government officials and public interest groups.

Appointing professionals with above-average technical expertise is obviously *insufficient* toward accomplishing a Project Manager's objectives. Project Manager expertise must be seen on a variety of levels and must include not only technical expertise but also administrative capability and a knowledge and understanding of customer wants and needs.

The typical project "liaison" role involves maintaining a variety of communication contacts. Social relationships through which information is transmitted must be developed and maintained. Project Managers must continually attend meetings, lunch with team members, and arrange to talk with influential experts and scientists on a regular basis. These contacts provide Project Managers with access to information unavailable to team members, thus strengthening their power and potential influence bases.

6.12 CONCLUSION

Simply appointing a professional with technical expertise to a Project Manager role is insufficient. Achieving a unity of effort within the complicated matrix project structure requires a firm appreciation of the problems and diverse needs facing a project team and an ability to create enthusiasm and commitment for the work itself. This ability evolves from a clear understanding of the nature of project leadership and the identification and development of a variety of influence and power bases.

REFERENCES

1. Hodgettes, R. M. "Leadership Techniques in the Project Organization," *Academy of Management Journal,* Vol. 11, 1968, pp. 211–219.
2. Thamhain, H. J., and Gemmill, G. R. "Influence Styles of Project Managers: Some Performance Correlates," *Academy of Management Journal,* Vol. 17, June 1974, pp. 216–224.
3. Gemmill, G., and Thamhaim, H., "The Effectiveness of Different Power Styles of Project Management in Gaining Support," *IEEE Transactions on Engineering Management,* Vol. EM-20, May 1973, pp. 38–44.
4. Gemmill, G., Thamhaim, H., and Wileman, D. L. "The Power Spectrum in Project Management," *Sloan Management Review,* Vol. 12, Fall 1970, pp. 15–25.
5. Ibid., p. 16.
6. Fiedler, F. "The Leadership Game; Matching Men to the Situation," *Organizational Dynamics,* Winter 1976, pp. 6–16.
7. Hershey, P., and Blanchard, K. H. *Management of Organizational Behavior Utilizing Group Resources,* 3rd ed. Englewood Cliffs, NJ: Prentice Hall, 1977.
8. Hershey, P. *The Situational Leader . . . The Other 59 Minutes,* Center for Leadership Studies, 1984 page 64.
9. French, J. R. P., and Raven B. "The Bases of Social Power," in D. Cartwright (Ed.), *Studies in Social Power.* Ann Arbor, MI: Institute for Social Research, The University of Michigan 1959, pp. 150–167.
10. Baher, B., and Wilemon, D. "A Summary of Major Research Findings Regarding the Human Element in Project Management." *Project Management Journal,* Vol. VIII, No. 1, March 1977, pp. 34–40.

Case Study*

A GIANT LEAP

Introduction

The steering committee was to meet in 15 minutes to decide the fate of the MP84 Project. "I don't know what the outcome of this meeting will be," lamented Pat Hall, a Department Supervisor in the Systems Development Laboratory. "I really thought that Stan and Dick could pull this project together. We really need that high speed modem, and now. . . ."

"Let's not quit before the final count," responded Donna Keyes, a Department Supervisor in the Software Engineering Laboratory. "Let's wait to see the outcome of this meeting and what the results of the steering committee's investigation have been. But it sure doesn't look like we'll ever get MP84 by 1984! I sure hope that we don't have to change the name to MP88! I have a gut feeling that if we had some more "management" and a little less salesmanship, we might be a lot closer to our goal. You know, I have seen project management at work, and I think that's what we need for this Project. Something has just got to reduce the slippage we have now."

Background History

VITAL, although a leader in the field of cable communication, had gotten off to a rather slow start entering the data communications market. By 1974, the growth rate of its major competitors in the field forced VITAL executives to do some serious thinking. It was predicted that advances in technology would result in the use of cable to transmit high speed data as well as television to homes and offices. But the competition was already stiff. It was, therefore, imperative that VITAL develop a modem that would be technologically superior to those of its competitors.

Marcia Freed, Engineering Manager of the Modem Division, and Sam Myers, Engineering Manager of the Product Development Division, were champions in the field of modem research and development. After many trials and investigations, Sam and Marcia firmly believed that they had the capabilities to employ state of the art technology and develop a 12 Mb/s modem to operate over 6MHz cable.

As most champions in the field tend to be, Sam and Marcia were young and enthusiastic, and they easily convinced top management of the merits of such a product. The MP84 was to employ every new technology available, such as a high speed equalizer, which was beyond the capability of their

---

* This case study was written for class discussion purposes only. It is not intended to reflect management practices or products of any particular company or organization.

competitors. A modem of the future was in reach and there wasn't a department or engineer that didn't want to get on the bandwagon, in spite of the high risk associated with it. Marcia and Sam agreed to deliver a Project Master Plan and Budget to their VP, Jerry Kane, within 30 days. Jerry Kane then convinced Bill Walter, President of VITAL, to fund the project.

Establishing Product Design

Having received the blessings of top management and funding for the project, Marcia, Sam, and their Section Heads, proceeded to develop a work breakdown structure (WBS), which outlined all major tasks involved in the development of the MP84 and all of the work necessary to accomplish each of these tasks. It was subsequent to the design of such a document that Marcia decided to call a meeting of the Engineering Managers whose engineers would perform the required tasks. Since the program was so extensive and required extremely diverse knowledge and technology, Marcia and Sam found themselves facing a group of Engineering Managers working for the Vice Presidents of Research, and Modem Technology, as well as Managers working in the Vice Presidential Area of Digital Signal Processing and Software. Marcia, once again, described the enthusiastic response of VITAL management; enthusiasm that was displayed in a budget of over 6 million dollars!

Marcia continued, "Ladies and Gentlemen, we've got carte blanche with this baby! Let's make it spin! As you can see from the work breakdown sheets you've each received, I've asked each of your Divisions to work on a specific portion of the project. I would like a detailed WBS from each of you, showing how you will break down your project into tasks, and the name of the Section Head accountable for each task. I'll also need a written quote of the budget you will require and a tentative schedule. I'd like everything within the week. My office will take your inputs and prepare the final WBS and budget.''

Off To A Flying Start

It wasn't surprising that each division involved responded to Marcia's initial request within the allotted week's time. No one wanted to be left out of this program with its almost unlimited budget and the professional recognition that came from designing a beyond-the-state-of-art modem. As a result of the enthusiasm displayed at the initial Engineering Manager meetings, Sam and Marcia created what they believed was to serve as the steering document of the MP84. The WBS was the official specifications document, designed to serve as a coordinating and steering mechanism for all the divisions involved.

Trouble Begins

But the MP84 spanned three Vice Presidential areas, three divisions, and five laboratories, each subcontracted to design a particular aspect of the modem or to provide a particular expertise. The program would eventually grow to involve 8 different departments. Few people, early in the life cycle of the program, truly understood the complexity and growth that lay ahead in the development of this product. Building on the newest of technologies, each functional laboratory that contributed to product development perceived the MP84 through its own functional perceptions. But with the absence of a central office, little control was exercised over design modifications, and as may be expected, design specifications changed at whim. While an initial planning document had been designed and used to parcel the work, it was very superficial. No coordination, communication or control existed between the laboratories. Parts designed were incompatible, deliveries were often late, and schedules mismatched. No mechanism had been established to deal with problems. There was little attempt at Systems Engineering and no true management of the program!

Although tremendous foresight and creativity were exercised in the conception of the new modem, little insight or understanding prevailed with regard to managing, scheduling, and controlling the operations involved. The problem of coordinating efforts between divisions just had not been considered.

It became increasingly obvious that the size and complexity of the product created enormous integration problems that the present traditional organization and management structure were not equipped to handle. A new organizational structure was needed.

A Project Organization is Formed

On March 14, 1984, Jerry Kane, Vice President of Data Communications, scheduled a meeting of the MP84 Steering Committee. The committee consisted of the Vice President of Research, Vice President of Broadcasting, Vice President of Planning, Vice President of Auditing, and the Vice President of Accounting. They had recently completed a review of the progress of the MP84, which consisted of a series of interviews with all Laboratory Heads involved in the program. They were dismayed to learn that the program was undergoing tremendous slippage. In fact, some of the key estimates used to plan the work program were off by as much as 600%.

> *Kane:* "Look guys, this program is a fiasco! It must be salvaged and to do that requires a whole new approach to management."
>
> *Vice President of Planning:* "What's the story here, Jerry? We've given your people every resources available to come up with this modem. Now you have a product that limps when we want to walk!"

Vice President of Auditing: "What's this 'Project Management' approach you've been advocating, Jerry? We'll try anything at this point!"

Kane: "It's simple, Ned. Although our engineers are talented and enthusiastic about their research, they're not team players. We've always managed to get the "pick of the litter," so to speak, what we need here now is not so much technical expertise as someone to pull this job together someone who'll provide the strength, direction, and leadership we need. What we really need is a *Project Manager!*"

Vice President of Planning: "And from what I understand of a systems approach, we need someone who is going to be objective enough to take all the variables into account."

Vice President of Accounting: "Not to mention cost! Your guys are spending money as if it's going out of style. Do you realize, how significantly productivity was underestimated? Some of our technical metrics for sizing this job were off by 600 per cent! There's been no up front budgeting, no limitations!"

Vice President of Planning: "Do you have a particular person in mind for this job, Jerry?"

Kane: "I've got just the person for the position. He's just what the program needs right now to pull it through this crisis. I think all of you know Hank Gale. Hank would be perfect for the job, and I think he will give us an honest assessment of exactly where we stand."

Vice President of Planning: "Tell us why you think he's our man."

Kane: "Hank has been with VITAL for over 20 years. He's an old-timer, with a great deal of experience managing projects for the military."

Vice President of Accounting: "I'll go along with the guy—provided he's conscious of costs. I also want one of my people, Barry Cohn, on his staff so that I know, on a regular basis, what we're paying for this chip."

Kane: "No problem there, Dennis. Hank isn't a 'primadonna,' and he's made it crystal-clear to me that the only way this is going to work is to put essential personnel—like Barry—full-time on the project. In addition, to make Project Management work, we must give him the authority to control the Program, so all of the Laboratory Heads must respond to Hank's request. I need your approval, Pete, and yours, Steve, since many of the Laboratory Heads work for you."

Vice President of Research: "Okay. If this is the only way to make a decent showing from this debacle, Pete and I will give him the backing he needs. You get him to pull this program together, and the sooner the better! Let's go to lunch."

There were many such meetings. Some for breakfast and lunch and others for dinner before the program was finally pulled together. The Vice Presidents strongly believed that they all worked too hard to allow the same problems to reoccur and agreed to meet the first Wednesday of each month at 11:00 A.M. to get a status report from Hank.

The Changeover Begins

It was a difficult decision for the Steering Committee to consider a new organizational structure, such as a matrix. But it's what Hank Gale insisted on—and he had their full support. "You're really going to have your work cut out for you, Hank," remarked Jerry Kane, Vice President of Computers and Software. "It's not going to be easy establishing a project management organizational structure on top of our traditional structure. We're going to have to absorb all the lumps and bruises and literally force the system.

Building a Project Team. The very first thing that Hank Gale did was to set up a Program Organization and a Program Office. On March 30, Hank approached his old friend and colleague Mike Sands. Hank described the problems confronting the project and asked Mike to assist him in managing the program. Mike was encouraged to identify his best engineers and to bring them on board. Mike turned to two friends, Donna Keyes and Pat Hall. "I'd like you two to join the Program Office team, which will consist of both of you, myself, and Barry Cohn of Accounting. Barry's got an excellent head on his shoulders. He's very interested in project management and will handle the scheduling and GANTT charts. He'll also be responsible for the control of funds and expenditures. In a sense he'll perform the functions of a comptroller and master scheduler. As far as our responsiblities go, Hank informs me that we are responsible for systems engineering and making sure that all parts are designed to fit together."

Implementing the Matrix Structure

In the early stages of the matrix implementation, few engineers working on the MP84 program understood matrix concepts fully. Most understood them hardly at all. However, believing that effective communication was central to the successful implementation of a matrix structure as well as to the accomplishment of the task at hand, Hank and his program team went about creating an atmosphere to support their beliefs. Mike, Donna, and Pat continually circulated among the group leaders, who provided the staff and technical resources needed by the Program Office. Asking questions, coordinating designs, and determining progress on the work packages, they worked hard toward gaining the trust and confidence of these first-level managers.

Mike, Donna, and Pat were required to study the tasks performed by each Laboratory and each Department Supervisor. To their dismay they found duplication of effort as well as uncoordinated yet highly dependent activities. Several group leaders were performing the same task, using different technologies. When their study was complete, they presented it to George.

It was early April when Hank called Mike, Donna, and Pat into his office.

I've studied your report. There's still havoc in our organization that results from a lack of information, untimely information, wrong data, or incomplete engineering. It's time to pull those reigns in tightly! The Engineers are probably going to resent it, but this program needs a lot more control. The system has gotten too big. You just can't run a large program with expenditures in the millions of dollars a year on the same informal way it operated in the past. There's a strong need for a more consistent definition of tasks and responsibilities. We've got to reorganize! Instead of 8 departments working on this program, I'm going to try to reduce redundancy and eliminate 3 of them and one Laboratory. One thing I must have is complete and timely information from each laboratory. And the only way I am going to get it is to *demand* it! I plan to hold weekly problem-solving meetings with all of our Laboratory Heads and I expect them to be well versed in the difficulties or problems, if any, facing the MP84 on their technical end of it. We'll put our heads together and hash it out. I want these team meetings to give people a chance to see and, hopefully, learn from each other's mistakes. And I'll lead each and every one of them! It's about time these managers were able to resolve their own differences and by the time this baby is brought home, they'll be experts in that state of the art. I want detailed and weekly status reports from anyone who has anything to do with this program. If there are any more impending slippages or delays, I want to know about them *before* they happen!

The Need for Control

Differences of opinion existed among the Laboratory Heads about the need for such highly structured procedures—differences that were not held quietly. Some of the Heads believed that procedures should merely be guides and could not realistically be captured on paper through a report or monitoring form. But in spite of the rather pervasive grumbling, the Laboratory Heads went along with Hank Gale—particularly since Hank had the support of their Vice Presidents and Engineering Managers.

One of the very first things Hank and his team did was to pull together one central development plan and an established set of design requirements. These documents were continuously updated such that they became the program's Bible.

Hank and his team also proceeded to establish what was refered to as a Change Control Board. This Board consisted of Mike, Donna, the rest of the program team, and *all* Laboratory Heads directly involved with the project.

Prior to any design changes, written specifications and requests for change were reviewed and signed off by all members of the board and circulated throughout the program team. Hank was going to make sure that there were no more surprises when it came to this product!

Hank held fast to his concept of monitoring procedures and weekly status reports. Subsequent planning sessions between Hank and his project team were done in great detail. By April 30, a detailed master plan and program schedule clarified the responsibilities and Contributions of the various organizations to the MP84 program. It seemed to include just about everything: A comprehensive listing of open issues to be addressed by Program and Project Planners in cooperation with Development Managers.

After the initial planning and design sessions, Hank continued to be updated on the program's progress and problems through his weekly Laboratory Head problem-solving meetings. And Hank was anything but understanding when it came to excuses from these Managers. In fact, he relied heavily on his hard-hitting, direct, and at times intimidating style to get the information he felt he truly needed to effectively manage a program of this size and magnitude.

Smoothing Out Relations

The Program team focused on directing and coordinating the program's work and made it their business to make frequent informal visits to each and every department. Never possessing official authority or direct control over project resources or personnel, Mike, Donna, and Pat had to rely on the strong working relationships they established with the Department Supervisors and Laboratory Heads, their common goal being that of success in the project. The Laboratory Heads were concerned about their performance at the weekly status meetings. So Mike, Donna, and Pat made sure to win their trust and support by preparing and distributing an agenda well in advance of each meeting and briefing each laboratory head as to the agenda and problems.

Before each meeting the program team always reported back to Hank Gale to brief him on all the problems he would encounter. Hank and the program team, knew that to produce the MP84 on time required that project milestones be kept and that slippage be eliminated. They also stressed the importance of communication and trust in the reconciliation of differences. The specialists were extremely anxious to do a good job, and the key was simply to get all parties working on the same problem. As Donna Keyes later explained, many misunderstandings developed over documentation. As Donna explained: "I had a case where I couldn't communicate to a group of design engineers what it was I needed for equipment specs. I tried but it was obvious that I wasn't getting across. So I took the design information available and tried to translate it into operating specs myself. With this in hand, I went back to them and showed it to them. They said, "Boy, that's pretty bad''; and they were right, it was pretty bad because it really wasn't detailed enough. But this gave them an opportunity to see what I wanted.''

QUESTIONS

1. What is Hank's style of leadership? Discuss why it is or is not appropriate for the project.

2. What factors help determine which leadership styles are appropriate for use?

3. What is Donna's style of leadership? Is it appropriate considering the environment in which it occurs?

4. How do the terms "authority," "power," and "influence" relate to this case?

5. What were the advantages of the matrix structure over the former functional approach? Disadvantages?

CHAPTER 7

CONFLICT IN THE PROJECT ENVIRONMENT [1]

7.1 INTRODUCTION

CONFLICT! It's one of those fascinating terms that evokes imagery heavily laden with negative connotations. Just think of some typical words or feelings that you associate with the term: anger, tension, animosity, battle, fear, hate, and distrust. We can go on and on. Webster's [2] dictionary defines conflict as "discouragement . . . , war, battle, collision . . . the opposition of persons." Conflict is often associated or equated with narrow mindedness, hostility, and win–lose situations. The connotation, and the dictionary definition, suggests that conflict is, indeed, a negative force; something to be avoided or, at best, suppressed. But in today's modern industrial environment conflict has a much greater, positive potential than its possible destruction or damage. If managed and approached effectively, conflict can be a vehicle for change, an integral part of problem-solving, and a catalyst that synergizes *diverse* ideas and *improved* relations.

Conflict, as we all well know, has the frightening potential of destroying a project, creating slippage, work stoppage, poor quality, and even sabotage. But conflict will exist on *all* projects. Because of the interdependent work of project specialists, the shared and limited resources, and the divergent technical opinions, goals, and organizational roles, conflict is an inevitable phenomenon. An important aspect of project management is the ability to effectively manage the conflict that arises from the rapidly changing, dynamic project environment.

If assessed and managed properly, conflict can actually achieve positive results for the project and project team. To achieve such results, Project and Functional Managers must recognize the many potential sources of conflict within the project organization, assess its impact on the project's near and long-term goals, and select, from a variety of resolution strategies, the most appropriate technique(s) for resolving the conflict and building a more cohesive project team.

The purpose of this chapter is to redefine conflict as a potentially positive vehicle for innovation and change, to explore and develop strategies that minimize or prevent its dysfunctional effects, and to help Project Managers to develop a variety of resolution techniques that help in building more productive and cohesive project teams.

7.2 THE BASIS OF CONFLICT

Organizations of the past were somewhat at an advantage when it came to minimizing conflict. Most production systems were established to produce standard goods through highly structured and repetitive procedures. Roles and relationships were clearly defined and decision-making authority placed in the hands of a select few. Top management defined the organization's goals, which employees tended to accept unquestionably for the good of the organization, and for their own job security and survival.

But times and markets have changed. To survive in today's dynamic and competitive environment, organizations have undergone significant transformations not only in structure but also in technology, economics, and the composition of the work force. High technology and automation, as well as dynamic, fragmented, and changing markets, force corporations to assess their traditional structures and create more fluid organizations capable of responding rapidly to a variety of internal and external demands. Decision-making is less centralized, as work groups and project teams are developed to generate and manage change. ''Integration of specializations'' has become the objective of the modern organization, and with it comes a series of incongruent, conflicting goals.

7.3 TRADITIONAL VERSUS CONTEMPORARY VIEWS OF CONFLICT

As indicated in Table 7.3-1, organizations and the managers that work within them tend to subscribe to one of two opposing views of conflict. The *traditional view* tends to see conflict as dysfunctional or harmful to the organization; something that should be avoided at all cost. According to this viewpoint, conflict is the result of either poor leadership or personality differences and is resolved by physical separation, or intervention by higher management. Traditional managers tend to either ignore or avoid conflict, or squelch or suppress it rapidly.

The *contemporary view* of conflict is more optimistic in nature. According to this view, conflict is an inevitable consequence of the complex organiza-

7.3-1   Traditional and Contemporary Views of Intergroup Conflict

Intergroup Conflict	Traditional View	Contemporary View
Philosophy	Conflict is dysfunctional to the organization and should be avoided	Conflict is an inevitable consequence of organizational interactions and can be resolved by identifying the sources of conflict; conflict can be a force for positive change in organizations
Cause	Conflict is caused by personality differences and a failure of leadership	Conflict is generally the result of the complexities of organizational systems
Resolution	Conflict is resolved by physical separation or the intervention by higher management levels	Conflict is resolved through identifying the causes and problem-solving

tional systems and human interactions. It is the predecessor of change and can actually be beneficial for a project if handled properly. Rather than suppress it, contemporary managers view conflict as an opportunity to confront and resolve the issues that prevent project success. By conducting problem-solving sessions to identify conflict causes, contemporary managers use conflict to improve group cohesion and increase project performance.

The contemporary view of conflict is a more positive and appropriate approach for a modern, high-technology organization. Conflict is seen as a natural and essential part of the problem-solving process. It is inevitable, especially in a matrix-project organization, where integration of specialization is the primary objective. Its value, however, depends on *how* project specialists regard and manage it.

Since conflict is inevitable, our goal is *not* to eliminate it, but to view it as an essential, healthy, and productive characteristic of project work. If managed properly conflict can enhance team productivity, stimulate innovation, and assure a quality product [3]. Through the constructive management and eventual resolution of conflict, it is possible to gain a broader understanding of the nature of the problem and its implications. By encouraging rather than suppressing the expression of conflicting opinions and ideas, managers can create a reservoir of alternatives from which a solution may eventually evolve. The excitement and energy that is generated from healthy divergent ideas, and the efforts made to resolve them, may actually help to blend a work *group* into a more cohesive *team*.

7.4 THE ANTECEDENTS OF CONFLICT

Although divergent ideas and disagreements may arise in all types of interactions, nine conditions have been identified that specifically predispose organizations toward conflict [4,5]:

1. Ambiguous roles, overlapping responsibilities.
2. Inconsistent and/or incompatible goals.
3. Communication barriers.
4. Interdependent tasks or activities.
5. Differentiation or specialization in organization.
6. Need for joint decision-making.
7. Need for consensus.
8. Procedures and regulations.
9. Unresolved, prior conflict.

These antecedent conditions are factors that are associated with prevalent organizational conflict. It is the degree to which a variety of these conditions exist in combination that creates the potential for conflict and determines its intensity within the organization and project team.

It is common, especially in the matrix structure, for two or more individuals, sections, departments, or divisions to have related and even *overlapping responsibilities* and *ambiguous roles*. Conflict in this situation is generated by the needs of the individual or the unit to define and establish its mission and purpose within the organization. *Incompatible goals* arise when departments or individuals must work interdependently yet perceive each other as having opposing missions. This is a common occurrence, for example, between Sales and Manufacturing. On one hand, sales frequently promises customization in order to obtain a small but lucrative order. On the other hand, manufacturing prefers high volume orders and may find the small, customized orders a nuisance because tooling mechanisms must be modified to produce limited quantities.

Communication difficulties create even further misunderstandings, as efforts to explain and negotiate the needs of the parties involved become blocked. Conflict further results when one discipline or function is dependent on the next for activities and resources. Modern, high-technology organizations are characterized by pools of specialists responsible for unique tasks. These *specialized groups* possess their own perspectives, language, and goals, which creates still another condition that can stimulate conflict. This situation intensifies as project teams work across the organiza-

Table 7.4-1   The Antecedents of Conflict

Factors Encouraging Cooperation	Factors Associated with Conflict
Common goals	Incompatible goals
Common values	Lack of common values
Similar backgrounds and perceptions	Disparate backgrounds and perceptions
Acceptable central authority	Weak, decentralized authority
Interdependent tasks	Interdependent tasks
Proximity in workplace	Physical or time separation
Consensus not necessary	Consensus necessary
Supervisor–subordinate expectations consistent	Supervisor–subordinate expectations inconsistent
Clear responsibilities and roles	Ambiguous responsibilities and roles
Limited structural differentiation	Extreme structural differentiation
Equitable work systems and workloads	Inequitable work systems and workloads
Adequate resources	Limited resources from a common pool
Noncompetitive relationships among members	Win–lose relationships among participants

tion to accomplish organizational goals. When groups of *divergent talents* must further reach consensus or agree among themselves, disagreements naturally occur and are difficult to manage. As *rules, procedures,* and *regulations,* natural parts of the project environment, restrict team member actions, team players may feel they are in opposition (or in "conflict") with the very organization they attempt to serve! Finally, unresolved conflicts tend *not* to dissipate but intensify and create a strained environment that precipitates even more intense and destructive conflicts.

Table 7.4-1 lists a variety of factors that serve as antecedents of conflict and cooperation.

7.5 THE DESTRUCTIVE ELEMENTS OF CONFLICT

Conflict is "destructive," "disruptive," or "distributive" when project team members fail to understand its value as a source of diverse ideas and alternatives or do not have or use constructive means, skills, or behaviors to channel the conflict into problem-solving discussions and deliberations.

Destructive conflict is frequently associated with competitive win–lose situations. It is characterized by a lack of team spirit and an "not-invented-here" (NIH), "get my own way" attitude. Group members "position" themselves as they stubbornly adhere to their own narrow viewpoints and fail to consider the possible value of another approach. Group decisions are stalemated, while members avoid the critical issues and engage in personal attacks. Negative attitudes and feelings grow more entrenched each day. Even if a decision is eventually reached, it is likely that few people will be pleased or satisfied. Destructive conflict, therefore, produces defensive and disruptive behavior. It is the type of conflict that gives "conflict" its bad name!

7.6 CONSTRUCTIVE POSITIVE APPROACHES

Constructive or integrative conflict arises from a climate characterized by open communication and mutual understanding. It develops only when project team members have acquired skills in effective communication and possess a variety of methods with which to handle the conflict. Members demonstrate a high sense of team spirit and recognize that disagreements evolve from a sincere commitment to the successful realization of project goals. Project Managers who model integrative conflict management strategies tend to be less apprehensive of disagreements and more willing to approach conflict rather than avoid it [6].

Integrative situations are characterized by supportive rather than defensive communication. Active empathetic listening prevails as team members grow to appreciate the merits of considering others' opinions and integrating a variety of viewpoints into the *best* possible solution. Group cohesiveness and trust evolve from the support and openness team members display toward one another. Team decisions are reached primarily by consensus and represent not merely a compilation but a *synergy* of the most positive aspects of the group's total solutions.

7.7 BUT DIFFERENCES ESCALATE!

As Figure 7.7-1 illustrates, however, even what appears to be the simplest of differences can escalate into destructive conflict.

Conflict may start as a simple difference of opinion over roles, procedures, or values, which may seem to be a simple misunderstanding of "I see

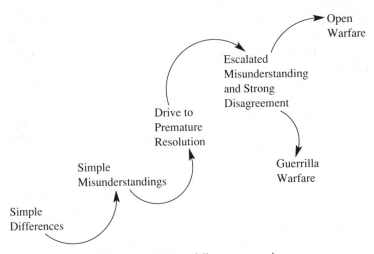

Figure 7.7-1 How differences escalate.

it my way, and you see it yours." Unfortunately, because many of us find opposition distasteful, we frequently react quite negatively and, at times, defensively in such situations. Feeling uncomfortable and perhaps even a little fearful of losing control of the situation, we tend to push to eliminate the conflict by *forcing* a solution through unilateral escalation or by establishing some virtually ineffective or watered-down approach. We do not allow ourselves the time to sufficiently determine where the actual differences lie. Thus, we permit the real conflict sources to grow stronger and more harmful. Participants grow entrenched in their positions, as they become more concerned over supporting their respective methods or procedures, than about the end result or common goal. This, in turn, leads to a state of open warfare, making it almost impossible to perceive or agree on a mutually beneficial result. Guerrilla warfare may further arise in the form of excessive project turnover, lack of commitment to project objectives, or even sabotage.

7.8 "A PROBLEM SITUATION"

Taking the time to actually examine the many sources of conflict is the first step toward dealing effectively in such situations. Unfortunately, finding the time to accomplish this objective is often a problem in itself. Let us examine a short and simple case situation.

Exercise: A Problem Situation

Ms. Payne, Supervisor of Product Sales, received a rush order from one of her company's most important customers. The customer requested a slight modification in one of the stock items. Ms. Payne was particularly concerned about this request since she is responsible for satisfying this influential customer. She received the call at 9:00 A.M.; the revised items had to be shipped by 5:00 P.M.

To save time, Ms. Payne went directly to the Manufacturing Supervisor, Mr. N. O. Tyme. Mr. N. O. Tyme wasted no words. Because he had "his own emergencies," he would not interrupt his work schedule to do this special job. Further, he complained about the continuous interruptions and requests from Ms. Payne's area.

Ms. Payne went to her manager, who, in turn, approached Mr. N. O. Tyme's manager. Mr. N. O. Tyme was ordered to put the rush job through. The special order was completed by early afternoon; however, Quality Control rejected the entire lot because of an extremely high error rate. The job was completed the next day—12 hours too late.

This event accentuated already existing differences between the Sales and Manufacturing areas.

1. What are the various sources of conflict?
2. How could conflict have been avoided or utilized?

For but a few moments, we have taken you out of the complicated high-tech project environment and into the "simpler" world of "widgets." Yet, many of us will easily identify with the laundry list of conflict sources this simple case generates.

It appears that one source of conflict may, indeed, be *personality differences*; that is, Payne and N. O. Tyme just do not like one another. Let us bear in mind, however, that what may *appear* to be personality conflict may, at times, actually represent *organizational role conflict*. This occurs when we ascribe and generalize negative characteristics to individuals who represent the various divisions within the organization and thus "personify" the organization itself (as represented by the following attitudes: "People from marketing can never understand engineering problems," "Support groups are incompetent," "Engineers make terrible marketers"). It also appears that *scheduling conflict* is occurring and that Payne and N. O. Tyme have a *difference of priorities*—Payne with her small but juicy orders and N.O. Tyme with his huge volume of widgets to produce. In addition, there appear to be *no procedures for special orders*. Or, as the phrase "to save time" may indicate, *procedures* for just this kind of situation may have been bypassed in order to expedite matters. *Lack of consultation* with Manufacturing certainly contributed to ill feelings and may have led to a situation of *premature commitment to the customer*. We advocate that saying "no" to a customer is unheard of. One never says "no" but "Indeed!" and then carefully informs the customer of the effect of the special request on schedules and cost. It is easy to see that the lack of negotiation on Payne's part, and the lack of "buy-in" from Manufacturing really did the job in, not to mention the hundreds of unresolved conflicts that have probably occurred prior to this incident.

We can go on and on. The point is that there is a wide variety of conflict sources in today's modern organizations and the first step toward working with conflict is to identify and understand its sources.

7.9 SOURCES OF CONFLICT IN THE PROJECT ORGANIZATION

Among the several forms of project organization, the one that breeds the greatest amount of conflict is the matrix organization. As discussed in detail in Chapter 6, the matrix Project Manager may perform the functions of planning, organizing, directing, and controlling but has little direct control over the human resources contributing to project objectives. Characteristics of the matrix environment, such as constant and rapid changes, shifting project personnel, and ambiguous and overlapping roles and responsibilities, all increase the potential for conflict.

Recognizing its primary sources is the first step toward managing conflict effectively. In fact, several studies [7–9] indicate that across many major, technology-oriented organizations, there appear to be several common sources of conflict that vary in intensity over the life cycle of the project.

Exercise

For the next few moments, think of your present or past project situation(s). List what you believe are the seven top sources of conflict and rank-order these sources according to their negative impact on project performance and morale. Assume that rank 1 has the greatest impact, while rank 7 has the least. Compare your rankings to that of the research discussed on the subsequent pages.

Rank	Source
1	_____
2	_____
3	_____
4	_____
5	_____
6	_____
7	_____

By now the message has become increasingly clear: Conflict *is* a natural phenomenon that occurs as a result of organizing work by complex, multidisciplinary project teams across some form of a matrix or hybrid structure. The unique conditions characteristic of matrix project environments, such as dual-boss relationships, complicated logistics and physical separation, communication barriers, scarce resources, and perceived incompatible goals or differing technical outlooks, all contribute to the potential for and the intensity of conflict.

If you are similar to the hundreds of Project Managers and engineers surveyed by Thamhaim and Wilemon in their classic 1975 study, your conflict sources may include some or all of the following fundamental areas:

Rank	Source
1	Schedules
2	Priorities
3	Staff
4	Technical opinions
5	Administrative procedures
6	Cost objectives and estimates
7	Personality

Table 7.9-1 provides a detailed explanation of each of these classic conflict sources.

Thamhain and Wilemon's work has been confirmed by a number of investigators working in a variety of project environments [10]. Most recently, Posner [11] attempted to replicate the classic study and examined the pattern of conflict intensity occurring today in a wide variety of technology-oriented organizations. Using survey instruments developed and validated in the earlier studies, respondents indicated the issues that were most likely to create conflict during a project and how the intensity of such sources varied over the life cycle.

Since much of the research concerning conflict was conducted over a decade ago, it is not surprising that the more recent 1986 study indicates the following changes in rank:

Table 7.9-1 Seven Classic Causes of Conflict in Project Management[a]

Potential Cause	Characteristics
Schedules	Disagreements that develop around the timing, sequencing, and scheduling of project-related tasks
Project priorities	Differing views by project participants over the importance of activities, tasks, and trade-offs that should be undertaken to achieve successful project completion
Personnel (staff)	Conflicts that arise around the staffing of the project team personnel from other functional and staff support areas from the desire to use another department's personnel for project support
Technical opinions and performance trade-offs	Disagreements that arise, particularly in technology-oriented projects, over technical issues, performance specifications, technical trade-offs, and the means to achieve performance
Administrative procedures	Managerial and administrative-oriented conflicts that develop over how the project will be managed; i.e., the definition of the Project Manager's reporting relationship, definition of responsibilities, interface relationships, project scope, operational requirements, plans of execution, negotiated work agreements with other groups, and procedures for administrative support
Cost	Conflict that develops over cost estimates from support areas regarding various project work breakdown packages; for example, the funds allocated to a functional support group might be perceived as insufficient for the support requested
Personality	Disagreements that tend to center on interpersonal differences rather than on "technical" issues; conflicts that are "ego-centered"

Adapted from Thamhain, H. J., and Wilemon, D. L. "Leadership, Conflict, and Program Management Effectiveness," *Sloan Management Review*, Vol. 19, No 1, Fall 1977, pp. 69–89.

1976	**1986**
Schedules	Schedules
Priorities	Costs
Staffing	Priorities
Technical Opinions	Staffing
Procedure	Technical Opinions
Costs	Personality
Personality	Procedures

As indicated, the pattern of conflict over costs differs markedly from the earlier studies, changing from sixth to second place. Conflict over administrative procedures, moreover, appears to have been a more intense issue a decade ago than in the more recent investigation, changing from fifth to final position in intensity.

A variety of circumstances exist that may explain the differences between the classic study and the more recent research. Differences over cost conflict may actually reflect the change from a U.S.-dominated to an intensely competitive worldwide market [11]. It may be further compounded by the changes in government contract pricing strategies, from a more flexible cost-plus basis to the more rigorous price-fixed approach. The diminished intensity of procedural conflict may further indicate the increased acceptance of project management strategies and related organizational forms.

7.10 LIFE-CYCLE CONFLICTS

Another interesting aspect of conflict in the project environment is not only that its sources can be consistently identified but also that the intensity of each source varies according to the specific phase within the life cycle.

Overall, the differences in conflict intensity do not differ significantly from one source to the next [11]. There are, however, specific issues or sources that dramatically change over the entire life cycle and impact on the productivity and success of project activities. This becomes evident as we examine each of the four generic life-cycle phases: *project formation, project build-up, main program,* and *installation.*

Table 7.10-1 presents and compares mean conflict intensity (illustrated by

Table 7.10-1 Mean Conflict Intensity over the Project Life Cycle

1976	1986
Schedules	Schedules
Priorities	Cost
Staffing	Priorities
Technical Issues	Staffing
Procedures	Technical Issues
Personality	Personality
Cost	Procedures

rank order) over the project life cycle reported by the classic 1975 Thamhain and Wilemon study as compared to Posner's [12] more recent findings.

Table 7.10-2 analyzes and compares the sources of conflict as they occur over each specific project life-cycle phase.

7.10.1 Project Conception

As Table 7.10-2 indicates, conflicts during the conception and early definition phase tend to occur regarding schedules, priorities, cost and staffing. Disagreements over schedules and priorities have remained consistently intense over the previous decade. In many instances, product development and production schedules have been established via market analyses, customer requests, and strategic planning decisions, with only minimal involvement or consultation from those project participants directly involved with implementing project plans. This lack of project participant input in determining the feasibility of the end date, compounded by the frustration of balancing ongoing projects, increases the likelihood of intense and destructive conflict occurring during the conception phase.

Obviously, the most significant change in conflict intensity over the last decade lies in the area(s) of cost and administrative procedures. Managing costs has become a central organizational issue today. As markets become increasingly fragmented and competitive, *project* costs are becoming as important a consideration as *product* costs. Moreover, as project teams are required to provide firm-fixed cost estimates to government and non-government customers, the accuracy of initial cost estimates becomes a critical determinant of profitability.

Table 7.10-2   Conflict Intensity over Each Phase of the Project Life Cycle

1976	1986
PROJECT FORMATION (conception and definition) Priorities Procedures Schedules Staffing Cost Technical Issues Personality	**PROJECT FORMATION** Schedules Cost Priorities Staffing Technical Issues Personality Procedures
PROJECT BUILDUP (design and development) Priorities Schedules Procedures Technical Issues Staffing Personality Cost	**PROJECT BUILDUP** Staffing Priorities Schedules Technical Issues Cost Personality Procedures
MAIN PROGRAM Schedules Technical issues Staffing Priorities Procedures, cost, and personality	**MAIN PROGRAM** Schedules Priorities Cost Technical issues Staffing Procedures
INSTALLATION Schedules Personality Staffing Priorities Cost Technical issues Procedures	**INSTALLATION** Schedules Cost Personality Priorities Staffing Procedures Technical issues

A decade ago, conflict regarding administrative procedures (during a project's conception phase) was more intense than it is today. This difference reflects an increased understanding and acceptance of project management procedures and techniques by private industry. What was primarily a U.S. Department of Defense planning, scheduling, and controlling mechanim has, within the past decade, been successfully modified and applied to smaller, commercial products and services.

7.10.2 Project Buildup

As projects progress through the buildup (design and early development) phase, disagreements regarding priorities and schedules continue to remain critical, while staffing issues grow in intensity. As organizations streamline operations to meet the demands of a highly cost competitive environment, limited resources become a central and highly volatile issue. Project engineers now attempt to implement project plans, designs, and specifications within the time frame determined in the previous phase. Conflict over costs, however, appears to decrease within the buildup phase of the project. Because this phase is generally characterized by design and testing, a project may not yet be sufficiently mature to cause any intense disagreements over actual project costs.

7.10.3 Main Program

As the project progresses from build-up to the main program (advanced prototype development and manufacturing), nearly *all* areas of conflict are likely to intensify. During this phase, meeting schedules becomes the most critical issue to successful project performance. While conflict over schedules has been ranked as having the greatest overall intensity across the life cycle, it reaches its peak during the main phase of the program. Integrating the various project support groups, which contribute to the main program effort, is a difficult task and frequently contributes to project slippage. Disagreements regarding technical issues further intensify especially when product designs have not been sufficiently "bought in" or committed to up front by manufacturing engineers. Personality conflict also intensifies, as pressures mount and efforts to bring the product to fruition increase.

7.10.4 Installation (Phaseout)

As the project nears completion, scheduling and cost continue as intense conflict issues, while priorities and technical issues lessen in intensity. Personality disagreements remain as one of the top three sources of conflict, as over a decade ago. As project activities wind down, project participants become increasingly concerned over future assignments. Feelings of abandonment and lack of appreciation may be experienced by current project staff as team players from previous phases move on to new project work. Cost concerns are a central issue, as money allocation continues to increase. Staffing concerns center around field and customer support, issues of system upkeep, and current engineering. Conflicts over priorities develop as new projects start up and take on greater organizational interest.

7.11 PLANNING FOR CONFLICT

One way to manage conflict is to wait for it to happen and then smother it with a barrage of tactical skills and interpersonal charm and talent. For this approach to be successful, project team members must be well versed in conflict resolution strategies and constantly on the alert for eruptions of dissonance.

Another route to managing conflict is through *preventive* planning. If Project Managers are aware of the intensity and impact of each potential conflict source, they may be in a better position to determine strategies to minimize their destructive effects and utilize conflict to encourage synergy and change. Mapping out sound project moves and making them at the right time will substantially reduce major crises. *Planning* is the key to keeping conflict at manageable levels.

Schedules, project priorities, and costs are the primary conflict sources, followed by resources and technical opinions. If project planning and scheduling functions are properly performed, the odds for meeting project control parameters are increased and conflict levels tend to diminish.

Communication barriers, conflicts of interest, and differences in managerial philosophies are also sources of conflict. If blended properly, project teams can be prepared to deal with conflicts on a routine basis. In managing conflict, project planning may prove to be as important as the plan itself. By involving participants in the planning process, personal commitment is generated.

Table 7.11-1 provides some specific strategies for minimizing the destructive effects in key conflict areas and identifies techniques for Project Managers to use for more effective project communication and control.

7.12 CONFLICT RESOLUTION STRATEGIES

The potential effects of conflict, whether detrimental or beneficial, will depend on the environment created by Project Managers as they manage the many conflict situations which arise.

There is no "correct" way to settle conflicts. Managing conflict situations requires flexibility, an awareness of the most *appropriate* resolution strategies available, and a knowledge of the impact they will have on the project team.

Table 7.11-1   High-Intensity Conflict Sources and Strategies for Minimizing Destructive Effects

Life-Cycle Phase	High-Intensity Conflict Source	Strategies
CONCEPTION	Priorities	Jointly define and establish a mission; define a master project plan; develop first two levels of a WBS; define customer needs and solicit input
	Cost	Generate preliminary product requirements; perform and communicate a detailed market analysis and study; determine initial cost estimates and requirements; determine resource and staffing allocation
	Schedules	Establish preliminary project schedule and fundamental hard milestones; solicit preliminary input from organizations involved; identify risk areas and ongoing projects; document and distribute schedule information
	Procedures	Establish project focal point and clearly delineate project administrative procedures; define roles and reporting relationships; establish appropriate project organization
PROJECT BUILDUP	Priorities	Provide feedback on previously established plans and feasibility; detail project scope and specifications; develop a detailed WBS with work packages; establish contingency plans
	Schedules	Establish regular status review meetings; provide for periodic design reviews; pinpoint hard milestones; utilize PERT; identify critical path; track progress

Table 7.11-1 *(Continued)*

Life-Cycle Phase	High-Intensity Conflict Source	Strategies
	Staffing	Identify resource allocation through a detailed WBS; provide ongoing feedback; clarify roles and relationships; establish a reponsibility chart
MAIN PROGRAM	Schedules	Track progress and update schedules; feed back schedule information on regular basis; reward accomplishments or major milestones; identify possible slippage areas early and take action (risk analysis)
	Priorities	Obtain early buy-in and consultation from main program engineers and participants; establish a Change Control Board
	Cost	Implement Earned Value Analysis (EVA); employ budgeting techniques that reflect life-cycle output needs
	Technical issues	Provide for frequent testing and integration; schedule regular project review meetings
INSTALLA-TION	Schedules	Detail installation, and customer training schedule; identify high-risk slippage areas; provide customer input to installation status
	Cost	Track and monitor ongoing project costs
	Personality	Maintain harmonious working relationships through team building approaches; train team in conflict management strategies; identify new project resource needs

Table 7.11-1 *(Continued)*

Life-Cycle Phase	High-Intensity Conflict Source	Strategies
	Staffing	Provide training for field and maintenance personnel, customer, etc.; identity and select Support Staff early

Basically, there are five common methods available for handling conflict [13]: *withdrawal, smoothing, compromise, forcing,* and *collaboration* (or integration). As is the case with any leadership function, the specific method of handling conflict will depend on a number of situational variables. The "best" approach, however, will be the one that removes the obstacles that block or prevent project completion, and works to build *better* project teams. Analyzing the alternatives and their possible effects places Project Managers in a better position to choose their strategies wisely.

Withdrawal (Denial/Retreating). The withdrawal approach is frequently associated with low degrees of conflict, or situations in which the conflict appears to have little impact on project activities. Here the Project Manager, or project team, chooses to ignore the conflict or deny that it even exists. As a "cooling-off" period or a method of "buying time," this strategy can be quite effective. Withdrawal usually is a short-term strategy that neither directly deals with the conflict at hand nor serves to blend a better or more cohesive team. Conflict tends *not* to evaporate, but to build in intensity. Withdrawal, therefore, is least appropriate when the issues are critical to project success.

Smoothing (Suppression). Sometimes, when individuals fail to recognize the constructive aspects of handling conflict openly, they may tend to suppress it by playing down differences in viewpoints or opinions and emphasizing the commonalities or strong points. The result may be superficial harmony. Smoothing may, however, serve as a clarifying tactic and may be used effectively to emphasize certain points to a colleague (or opponent) prior to entering negotiations. In this sense, the typical pep-talk may be regarded as a smoothing tactic.

Smoothing over the conflict, however, does not help to eliminate it and is, of course, an inappropriate approach to use when others are ready and willing to deal with the issue. If the issues are relatively minor, however, and the time for problem-solving is limited, smoothing or suppression may be the most appropriate action.

Forcing (Power or Dominance). The use of power position or perhaps even majority rule is frequently used to force or "persuade" participants to reach a decision. It usually involves exerting one's point of view at the expense of another. In many instances, the use of forcing results in antagonism and resentment, which can increase conflicts. Forcing is a method that is most appropriate when used as a last resort.

Compromise (Negotiation). Compromise ("You give a little and I'll give a little") involves considering the various issues and determining a "middle-of-the-road" solution. Although valued and rewarded in our society as a conflict resolution strategy, compromising often leads to watered-down and ineffective solutions in which commitment by all parties may be dubious. Compromise is a situation in which neither party can win but that generates some "acceptable" form of resolution. It becomes a somewhat risky resolution strategy when considering disagreements over quality or technical performance.

Collaboration (Integration). The collaborative conflict resolution strategy involves recognition of the positive aspects of the conflict in terms of generating a variety of alternatives and solutions with regard to the specific problem at hand. Unlike compromise, where parties enter into negotiation prepared to experience some degree of loss, those involved in a collaborative effort fully expect to modify their original view, as the group's work progresses. Emphasis is placed on confronting the conflict, determining common goals, and generating a group solution. Often referred to as a "confronting" strategy, the assumption underlying this integrative approach is that the value of the group effort exceeds the sum of each individual's contribution. The result of a collaborative effort is a win–win situation for all. Collaboration becomes even more difficult, however, if time, consensus-seeking skills, and commitment are not present.

Choosing an appropriate conflict strategy is not an easy task. Conditions in which conflict occurs vary and, therefore, demand a sensitive yet objective eye. As Table 7.12-1 indicates, effective management of conflict requires much more than selecting one's best "personal" style and eradicating the differences among opposing parties. It involves recognizing and investigating the various sources of conflict that exist and then determining and predicting the *effects* of a resolution strategy on the project team's performance. This may involve using a variety of conflict resolution strategies.

7.13 THE "BEST" CONFLICT-SOLVING STRATEGY

Of the five basic resolution strategies, the collaborative–integrative approach appears to be the ideal method for resolving conflict. The collabora-

Table 7.12-1   Using Conflict Resolution Stategies

Methods	What Happens When Used:	Appropriate to Use When:	Inappropriate to Use When:
Denial or Withdrawal	Person tries to solve problem by denying its existence. Results in win/lose.	Issue is relatively unimportant; timing is wrong; cooling off period is needed; short-term use.	Issue is important; when issue will not disappear, but build.
Suppression or Smoothing Over	Differences are played down; surface harmony exists. Results in win/lose in forms of resentment, defensiveness, and possible sabotage if issue remains suppressed.	Same as above, also when . preservation of relationship is more important at the moment.	Reluctance to deal with conflict leads to evasion of an important issue; when others are ready and willing to deal with issue.
Power or Dominance	One's authority, position, majority rule, or a persuasive minority settles the conflict. Results in win/lose if the dominated party sees no hope for self.	When power comes with position of authority; when this method has been agreed upon.	Losers have no way to express needs; could result in future disruptions.
Compromise or Negotiation	Each party gives up something in order to meet midway. Results in win/lose if "middle of the road" position ignores the real diversity of the issue.	Both parties have enough leeway to give; resources are limited; when win/lose stance is undesirable.	Original inflated position is unrealistic; solution is watered down to be *effective*; commitment is doubted by parties involved.

Table 7.12-1 *(Continued)*

Methods	What Happens When Used:	Appropriate to Use When:	Inappropriate to Use When:
Collaboration	Abilities, values, and expertise of all are recognized; each person's position is clear, but emphasis is on group solution. Results in win/win for all.	Time is available to complete the process; parties are committed and trained in use of process.	The conditions of time, abilities, and commitment are not present.

tive strategy is preferred because it assumes that all parties involved will come out winners. In fact, a variety of investigations indicate that although Project Managers used the full spectrum of conflict handling modes in managing diverse conflict situations, they most frequently relied on collaborative techniques. Compromise was ranked as second preferred, with smoothing ranked third, followed by forcing and withdrawal [14,15].

From a management perspective, it is impossible to use collaborative strategies at all times and for all conflict situations. The choice of an appropriate conflict resolution strategy, in fact, relies on a number of variables, which include:

- Power position of the conflicting parties.
- Managerial and personal philosophy.
- Impact of the conflict situation on the project's schedule.
- Bottom-line monetary impact.
- Team building effects.

Project Managers who hold powerful positions within the organization favor both collaboration and forcing resolution strategies when the negative consequences of the conflict are high, but may tend to use smoothing, withdrawal, and compromise when the stakes are low.

Project Managers with lower organizational power positions, such as those frequently found in matrix organizations, tend to use collaboration and compromise for high-risk conflict while substituting compromise strategies for withdrawal when the effects of the conflict were considered low [16].

It appears, therefore, that there is no one best conflict resolution strategy. Smoothing works well to clarify certain issues and to stress common goals, while withdrawal may be appropriate when one does not have all the facts and needs to delay making a decision. Compromise is best used when both parties can afford to give something up. And forcing, although a win–lose situation, may be necessary when the conflict has reached a crisis phase. Integrative collaboration, although lengthy and requiring skill at consensus seeking, actually synthesizes *all* approaches and should be used as the preferred mode when time and skill is available.

Exercise

Using Conflict Resolution Strategies: Like effective leadership, conflict resolution strategies depend on a number of variables or conditions that help to determine the most "appropriate" course of action. There is no one best conflict resolution style. Power positions, the impact of the strategy on project team players, and where you are in the life cycle of the project may be included among the determining factors.

On the basis of your knowledge of conflict resolution strategies and their appropriateness for certain situations, complete the sentences below. Try to be as specific as possible regarding the characteristics of the situation that make this conflict management style suitable.

Smoothing or suppressing makes sense when/if_____

Avoiding or withdrawing makes sense when/if_____

Forcing or dominating makes sense when/if_____

Compromising makes sense when/if_____

Collaborating or integrating makes sense when/if_____

7.14 KEY STEPS TO MANAGING CONFLICT

Effective Project Managers realize that conflict is inevitable and, therefore, develop procedures and techniques for minimizing its negative effects and maximizing its constructive potential. These may include:

- Analyzing the problem in terms of the variety of situations that lead to conflict.
- Assessing the effect of a particular approach or methodology on the conflict and the project team.
- Developing the appropriate atmosphere or conditions for negotiation and resolution.

Should a conflict occur within a project team, a confrontation meeting is necessary between conflicting departments or parties. The Project Manager should then be aware of the recommended actions and sequence of events.

These include the following key steps:

Key Steps	Application Ideas
1. Ensure that key people affected by the conflict are involved	—
2. State the desired outcome of the situation	State the status quo of the conflict in terms of available facts, and point out the impact on the organization and the team
3. Have members of the group describe their view of the situation	Hidden issues or agendas may surface Avoid discussion of solutions at this time Avoid arguing Probe for what facts have led to a particular point of view, but do not criticize a member of the group for taking a position Seek additional viewpoints, if necessary
4. State the problem as understood by members of the group	Most problems are a product of misunderstanding; often, stating the problems clarifies the issue
5. Speculate on possible solution	Be creative; encourage others to come up with unique ideas Allow people to contribute freely and openly Deal with ideas, not personalities Build on points of agreement and reconciliation Focus on the overall goals of the organization Allow ideas to develop without prejudgment

	Do not let absence of data lead the group to reject concepts that might be valuable
	Develop a list of possible solutions
6. Evaluate ideas in terms of objective(s)	Decide how ideas would contribute to solutions and to superordinate goals
	Follow ideas through to their consequences and ultimate impact on people, costs and systems
	Seek additional data to validate ideas
7. Negotiate a resolution; test for clarity and agreement	Conflicts are resolved through flexibility and creativity. Each group member should clearly see how the solution would benefit the individual, the team and the organization
8. If the problem cannot be resolved in the group, seek mediation from a third party	Sometimes conflicts are reduced to contests because of negativity and inflexibility; in that case, a third party should decide the merits of each side's point of view and help make a final determination

Developing expertise in project management strategies will help a Project Manager to minimize conflict throughout the project life cycle. Recommendations for improving Project Manager effectiveness and minimizing conflict include:

1. Communicating key decisions in a timely fashion to project related personnel.
2. Adapting leadership style to the status of the project and the needs of project specialists.
3. Recognizing the primary determinates of conflict, when they are likely to occur over the life cycle, and the effectiveness of conflict-handling approaches.
4. Experimenting with alternative conflict-handling modes.
5. Providing work challenge to motivate support staff.
6. Developing and maintaining technical expertise.
7. Planning early and effectively in the life cycle.
8. Demonstrating concern for project team members.

Every effort should be made to integrate, right at the start, the various functional groups affected by the project. Product development teams should include at least one person from Marketing, Engineering, and Manufacturing. Letting people know what life will be like after the project is complete helps to keep them on track and headed for the goal. High-stress environments should be loosened up by "hokey," but effective team blending activities, prizes, functions, and parties. One top 500 corporation, during a major product development effort, would celebrate hard milestones by renting out football stadiums for team family picnics. Each team player had the opportunity to run through the goal posts as the scoreboard flashed their name to the jubilant crowd. Finally, since research consistently predicts what we can expect, resolving conflicts early will prevent slippage and ill feelings later on in the project's life cycle.

7.15 CONCLUSION

Conflict is an inevitable and necessary part of the project environment. Given the proper atmosphere, attitudes, and training, however, disagreements can actually broaden perspectives and stimulate innovative, productive and cohesive interactions. In their effort to deal with the uncertainty of situations such as changing technologies, dynamic competitive markets, and unstable economic conditions, managers have found that resulting conflicts can actually become predecessors to change. Project Managers who realize that preventing conflicts is as important as solving them, are likely to be effective.

REFERENCES

1. Kezsbom, D. S. "Creating an Effective Project Team: Dealing with Conflict," *IEEE Communications Magazine,* Vol. 21, No. 1, January 1983, pp. 54–55.
2. *Webster's Seventh New Collegiate Dictionary.* Springfield, MA: G&C Merriman, 1967.
3. Hill, R. "Managing Interpersonal Conflict in Project Teams," *Sloan Management Review,* Vol. 19, No. 2, 1977, pp. 45–61.
4. Filley, A. C. *Interpersonal Conflict Resolution.* Glenview, IL: Scott Foresman and Company, 1975.
5. Kirchof, K. S., and Adams, J. R. *Conflict Management for Project Managers.* Drexel Hill, PA: Project Management Institute, 1982. Drexel Hill, Pennsylvania.
6. Hill, R. "Managing Interpersonal Conflict in Project Teams," Sloan Management Review Vol. 19, No. 2, 1977, pp. 45–61.

7. Thamhain, H. J., and Wilemon, D. L. "Conflict Management in Project Life Cycles," *Sloan Management Review,* Vol. 17, No. 3, Summer 1975, pp. 31–50.

8. Thamhain, H. J., and Wilemon, D. L. "Leadership, Conflict and Program Management Effectiveness," *Sloan Management Review,* Vol. 19, No. 1, Fall 1977, pp. 69–89.

9. Posner, B. "What's All the Fighting About? Conflicts in Project Management," *IEEE Transactions on Engineering Management,* Vol. EH-33, No. 4, November 1986, pp. 207–211.

10. Eschman, and Lee, "Conflict in Civilian and Air Force Program/Project Organizations: A Comparative Study," September 1977, LSSR3-77B, A047230, p. 168.

11. Posner, B. "What's All the Fighting About? Conflict in Project Management," *IEEE Transactions on Engineering Management,* Vol. EM-33, No. 4, November 1986, pp 207–211.

12. Ibid.

13. Blake, R., Mouton, J. S. *The Managerial Grid.* Houston, TX: Gulf Publishing, 1964.

14. Thamhain, H. J., and Wilemon, D. L. "Conflict Management in Project Life Cycles," *Sloan Management Review,* Vol. 17, No. 3, Summer 1975, pp. 31–50; "Leadership, Conflict and Program Management Effectiveness," *Sloan Management Review,* Vol. 19, No. 1, Fall 1977, pp. 69–89.

15. Posner, B. "What's All the Fighting About? Conflicts in Project Management," *IEEE Transactions on Engineering Management,* Vol. EM-33, No. 4, November 1986, pp. 207–211.

16. Thamhain, H. J., and Wilemon, D. L. "Conflict Management in Project Life Cycles," *Sloan Management Review,* Vol. 17, No. 3, Summer 1975, pp. 31–50; "Leadership, Conflict and Program Management Effectiveness," *Sloan Management Review,* Vol. 19, No. 1, Fall 1977, pp. 69–89.

Case Study

INTERNATIONAL ELECTRONICS—A CONFLICT SIMULATION EXERCISE

International electronics (IE) is a medium-size corporation specializing in the design and manufacturing of communications systems and components. Located within the Northeastern United States, IE has experienced a relatively high growth rate within the last decade.

Recent market changes, however, have had a negative impact on IE's bottom line. Within the last 16 months profits have dropped by 10%, as IE products entered a highly fragmented and competitive market.

Last year, a project team was established in order to coordinate the introduction of the new PX1090 Communications System, to be used in the aerospace industry and in government defense. According to detailed market analyses, the new venture has tremendous potential of capturing quite a lucrative market for International Electronics. You were pleased to accept the position of Project Manager for this important mission. You report to the Vice President of Engineering Systems who, in turn, reports to the Divisional Vice President.

This morning, 12 months into the 26-month project, you receive a memo from the newly appointed Manager of Development. As part of his first official duties, he informs you that he has changed the RF input stage of the system, formerly analog, to a newer digital design. The memo also contains somewhat sketchy but supportive documentation as to the increased reliability of the product.

Changing the system from analog to digital will require major revisions and will set back total system implementation by at least 1 year. You are also aware that a considerable amount of time and research went into the design of this product, and how important timely delivery will be to future contracts and company growth.

Unhappy over the Development Manager's memo, you decide to walk in on him and discuss the situation and try to get him to hold off on his change request. The discussion becomes increasingly entrenched as he stresses the benefits of his plan and the errors and lack of information on which the previous plan was based. You, however, point out the benefits of placing this product in the hands of the user as soon as possible. Each of you becomes increasingly angry, and the meeting ends on a less than positive note.

At this point, you are extremely upset and decide that something must be done to ensure the success of the project. You see the alternatives as follows:

A. You can send a memo to the Development Manager, explaining your position and how the present design will make the company the foremost producer of the system in the world.

B. You can ask the Development Manager to meet with you for a full day next week in order to work out your differences and come up with an alternative solution.

C. Let it go for now; he will probably cool off soon enough and the crisis will be over.

D. You can go to the Divisional Vice President and request that former requirements be adhered to.

E. You can invite the Divisional Vice President to the next meeting to stress the importance of the project to the company's future.

F. You can march right back into the Development Manager's office and ask him to justify his position.

G. You can send a letter to the Divisional Vice President, resigning your position on the project team.

H. You can invite a representative from the Divisional Vice President's office to arbitrate all team problems or issues.

I. Send a letter to the Development Manager (with copies to all team members, the Divisional Vice President, and the President) indicating that opposition is holding up a potentially profitable project.

J. Ask the other department managers, who may realize a schedule delay, to convince the Development Manager to ease his request or incorporate it in the spin-off series.

Rank-order your alternatives and *identify* the resolution strategy you believe it represents (i.e., smoothing, withdrawal, forcing, compromising, collboration).

Discussion

As we discussed earlier, there is never any "correct" conflict resolution strategy. In this exercise, the various alternatives represent strategies that must be viewed from the perspective of the specific life cycle phase, the intensity of the conflict, the nature of the relationships of the individuals or departments involved, and the long- as well as short-term consequences of implementing the alterantive.

Let's examine each of the alternatives presented in terms of their *effect on the project team* and *how they address or resolve the conflict at hand*:

A. This conflict resolution strategy represents a *smoothing* or clarifying strategy. By sending a memo and stating your position, you are explaining your motivations and objectives and how they match the project's needs. The purpose of this strategy would be to provide the Development Manager with a greater perspective and understanding of the problem at hand, and an opportunity to reevaluate any previous decision. It may be wise to follow a smoothing or clarifying approach with subsequent collaboration. The danger of this strategy is that it does not directly deal with all the conflict issues at hand, namely, the Development Manager's objections.

B. The conflict resolution strategy here is *collaboration* or *integration*. By suggesting to meet for a full day, the Project Manager has indicated commitment to integrating his/her ideas with those of the Development Manager in the hope of generating a brand new solution. The emphasis here is on generating a solution to the problem through sharing expertise and ability. The outcome, therefore, is most likely a *win-win* solution to the conflict and will create the most positive long-term solution for the project and project team. Unfortunately, such an approach also takes the greatest amount of time. This solution, naturally, is the *most favored*, provided time and skills are available.

C. "Letting go" represents a form of *withdrawal* or *retreating* and may be extremely appropriate as a *short-term* "cooling-off" solution. Ignoring or denying the conflict, however, will not make it go away. It is best to follow through on such a cooling-off period with more constructive integrative, problem-solving approaches. Conflict, as you recall, intensifies over time.

D. This is a *forcing* strategy as it is a direct request to use power and position to resolve the conflict unilaterally. If both managers, however, had jointly agreed to approach the Divisional Vice President for clarification of organizational objectives, we would consider the strategy to be one of escalation and in the long run more desirable than forcing. Forcing is frequently used by Project Managers in the face of scheduling crises; however, if used to excess, the strategy may fail to blend team members and lead to long-run disruptive feelings of resentment.

E. As indicated previously, conflict resolution strategies, like leadership, are situational and depend on our attitutes, power position, project life-cycle phase, time, and other factors. This alternative can actually be viewed as two strategies. On one hand, the pep-talk that the Divisional Vice President delivers may be seen as a way of clarifying the importance of the project to team players and the need to reconsider the issues that are creating conflict. Such a *smoothing* strategy is represented by the Vice President's attempt to play down the differences among factions and stress the common goal. Another way to view this strategy is as a subtle form of *forcing*. Since it was the Project Manager who invited the Vice President to the meeting, it may be the Project Manager's desire to create an opportunity to be the one to discuss the issues and present the dangers of the conflict situation *unilaterally* to the Divisional Vice President. This strategy, of course, could result in resentment and hostility on the part of the other team members should they feel manipulated into submission.

F. This *confrontational* approach in some cases may place the Development Manager in a somewhat defensive position and may entrench the conflict further. However, confrontation of this nature may be successful depending on the previous relations of the involved parties. We all know individuals (and couples) who are capable of thrashing it out until a solution is finally generated. If the Project and Development Managers have had a strong relationship in the past and have had positive experiences in which conflict was resolved in this manner, the approach may lead to the beneficial win–win solution. Immediately confronting a heated situation, however, should be considered with care!

G. Resigning from the project team, ostensibly, represents *withdrawal* and is the least preferred method as viewed by most Project Managers. However, depending on the power position of the Project Manager within the organization, Alternative G may be viewed as a *forcing* mechanism. If the Project Manager's responsibilities and duties cannot at this time be adequately accomplished by someone else, and if the project is in a critical life-cycle situation, this strategy may be a powerful, although dangerous, tool to accomplish unilateral objectives.

H. Again, alternative H can be viewed as a combination of conflict-resolution strategies, depending on the rationale behind the approach. Arbitration, for one, is a form of *compromise*. The arbitrator with the assistance of the opposing parties will seek to find some middle-of-the-road solution and generate some resolution to the conflict at hand. However, by inviting a representative from the Divisional Vice President's staff to arbitrate *all* issues and problems, you have unilaterally relinquished all decision-making authority and, in effect, have *withdrawn* from a position of independence. Since this decision was made without the negotiation and acceptance of all concerned, lack of respect may develop for the Project Manager.

I. Obviously, this strategy is *forcing*. What we usually recommend in our seminars is, if you implement Alternative I, then you really have to perform G (resignation). This alternative is certainly a dangerous strategy, depending on where the project is in the life cycle, for it will serve to polarize and destroy a project team. Writing such a letter with carbon copies to all is an aggressive strategy because of the degree of forcing that is practiced. Although sending a letter might serve to clarify certain issues, in this instance it is not the content of the letter that is significant but the fact that pressure will be exerted on the Development Manager from *all* organizational directions. Politically, it is a very dangerous strategy and should be avoided at all costs.

J. What a difference a well-thought-out plan makes! Although a *forcing* mechanism, this strategy is one of the preferred techniques of *using power peacefully* to one's advantage. By lobbying with the other Functional Managers on the team, the Project Manager has additional power to exert on the opposing manager. At the same time, incorporating the Development Manager's plan in a spin-off series, provides that person the recognition needed and the corporation with a new product line!

Now review your ranking in terms of the categories of conflict resolution. If you are like most Project Managers in a matrix environment, your conflict resolution strategies will be from most preferred to least preferred:

- Collaboration
- Compromise
- Smoothing
- Forcing
- Withdrawal

CHAPTER 8

COMMUNICATION IN THE PROJECT ENVIRONMENT [1,2]

8.1 INTRODUCTION

Proper communication is essential in any organization or interpersonal endeavor. Project work, however, is particularly susceptible to communication difficulties because of the unique characteristics of projects and of the matrix organization in which they are frequently found. Overlapping responsibilities, decentralized decision-making processes, complex interface points, and the tremendous potential for conflict all place a strain on the processes of communication. Yet, in spite of all its related difficulties, communication is the largest single factor determining the quality, efficiency, satisfaction and productivity of a project team [3].

One critical role of a Project Manager is that of "communicator." The extent to which Project Managers can effectively obtain and disseminate accurate information directly affects team performance in terms of coordination and integration of effort. Figure 8.1-1 depicts the Project Manager's role as "communicator" within the project structure.

It is essential to project success that the Project Manager be a communicator to senior management, team specialists, and individuals outside the project who have an interest in the project's result. Assuring effective communication between these important groups rests largely in the hands of the Project Manager. Unless pertinent information is disseminated to the *right* people, at the *proper* time and through an *appropriate* manner, the project's performance may suffer. The Project Manager who fails to decipher and pass on appropriate information in a timely manner may personally become the bottleneck of the project and cause its demise.

Project Managers, therefore, must expend considerable effort in learning to communicate effectively. This chapter provides a summary of fundamen-

243

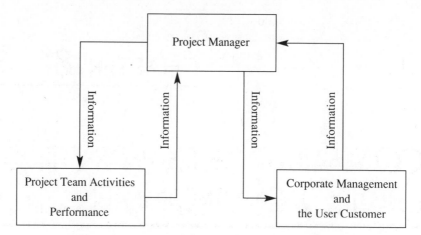

Figure 8.1-1 The project manager's communicator role.

tal communications theory and presents strategies that lead to a more efficient and productive project environment.

8.2 DOCUMENTATION VERSUS COMMUNICATION: COMMUNICATION DEFINED

Communication has been defined several ways. Many of our readers are probably aware that there are several sciences of communication: kinesics (body language), proxemics (distance or spatial language), linguistics, theater, graphic arts, and others, most of which are beyond the scope of this chapter. Our objective is to pinpoint a process that assists in creating more effective project communication. Some definitions include:

The exchange of facts, thoughts, opinions, or emotions.
The ability to transmit ideas, thoughts, or commands to the degree to which the other person understands or knows concisely what you are transmitting.
Getting the right information to the right place at the right time and in the correct manner, using various methods of feedback.
A two-way transfer of ideas and information through an acceptable medium to establish a mutual understanding.

One of the most inclusive of definitions is: "*Communication is any behavior that results in an exchange of meaning*" [4]. This encompasses what is said and what is not said, what is done and what is not done, how it is said, and how it is done. Through this definition it becomes apparent that everyone communicates, although not everyone communicates effectively.

Communicating with fellow project engineers or technical specialists usually poses few problems. Because engineering is a science based on precise symbols and use of equations, technical jargon helps to convey, in a precise fashion, the difficulties involved with the system. Unfortunately, spoken symbols used in *interpersonal communications* are not as precise in meaning as the symbols used in engineering or in technical project communication. In most cases, technical judgments and decisions must be presented and defined through a complicated verbal communication process. If the needs of the project system cannot be clearly conveyed to contractors and colleagues alike, costly errors may result.

Verbal communication, or the ability to talk, is based on a common system of spoken symbols that are used to relay *both ideas* and *emotions*. It is a vital and powerful tool for the passage of information, the delivery of explanations, and the imparting of instructions. It is the vehicle by which a negotiator gleans information and develops an understanding of the problems or motives of the other negotiator.

Verbal communication has many advantages. The choice of words, for example, and the clarity of expression can be carefully combined to strengthen meaning and impact. Field studies [5,6] indicate that engineers and technical specialists spend between 50 and 75% of their time communicating verbally with others. Verbal communication can be an efficient medium for the transfer of information. It permits timely exchange, rapid feedback, and the immediate synthesis of information.

Verbal communication, however, has disadvantages. As the complexity of the project increases, for example, the ability to transmit precise verbal descriptions diminishes rapidly. Jargon develops, which affords greater verbal accuracy among technical people but limits the range of understanding beyond the technical community. Since project members must interface with contributing lines of business, top management, marketing, and the customer, they must understand not only their own form of communication, but that of their project counterparts.

Effective communication skills are not something that can be achieved in a classroom setting alone. It requires a mix of concepts, practices, and skills, supplemented by determined effort, conscious practice, and firm resolve.

8.3 UNDERSTANDING THE COMMUNICATION PROCESS

As Figure 8.3-1 indicates, effective communication is not as simple as merely moving from point *A* to point *B*. Project communication features a series of distinct, yet interactive, and at times, even overlapping events. These events, in turn, are subject to a variety of influences that hinder or block effective communication.

The basic components of the communication model include a source, an encoder, a message, a channel, a decoder, a receiver, feedback, and noise. Table 8.3-1 outlines these elements. Communication, naturally, originates at

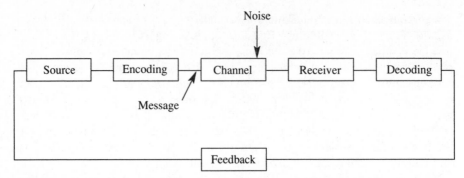

Figure 8.3-1 A communications model. Adapted from James L. Gibson, John M. Ivancevich, and James H. Donnelly, Jr., *Organizations, Structure, Processes, Behavior,* Business Publications, Dallas, Texas, 1973, p. 166.

the *source:* some member of the project team who has information to transmit to someone either internal or external to the organization. The process may be triggered by an idea or a concept that is, perhaps, unique and has never yet been communicated. This is frequently the case in basic research or in the development of a new methodology. Because of the newness of the idea, proposal, or concept, it may lack clarity or be incomplete.

In spite of these obstacles, the idea we conceive is next *encoded,* or translated into some language or set of "symbols" that, hopefully, expresses what the source wishes to convey. These "symbols" are influenced by the profession, education, culture, and values of the source, or sender, as that person digs within the realm of personal experiences for the appropriate language or code.

The purpose or intent of the source is expressed in the form of a *message.* This description must be relayed to the receiver through a *medium* or a *channel.* In a project organization this may be face-to-face communication, as in meetings or presentations, a telephone call, or written communication, such as technical publications, documentation, or test plans. *Noise* on the channel is any occurrence that causes a breakdown, interference, or distraction within the communication process. As in electronics communication, noise in interpersonal communication is anything that distorts the signal. In the "people sense," noise is anything that reduces the clarity and fidelity of the message, such as misunderstanding and misinterpretation based on the preconceived value judgments of both the sender and the receiver.

Finally, it is now up to the *receiver* to pick up the message and *decode* or interpret it in light of personal life experiences, value system, or frame of reference. The *feedback* the receiver provides the source provides insight

Table 8.3-1   Elements of the Communication Model

Source: some member of the organization who has information to communicate or convey to someone else in the organization

Encoder: an encoder process involves the translation of the original ideas into a set of symbols that express what the source wishes to communicate

Message: the actual physical product of the encoding process; the message is, basically, what the source hopes to communicate

Channel: the medium through which the message is carried from the source to the receiver; in an organization this may be face-to-face communication, a telephone call, word of mouth, group meetings, etc.

Decoder–receiver: provides for the process of communication to be completed; the message (sent by the source) must be decoded by the receiver–recipient in terms of a schema or meaning; the recipient will interpret the message in light of personal life experiences, value system, frame of reference, etc.

Feedback: feedback provides insight into the degree of fidelity that the message has—message fidelity is rarely perfect; feedback gives the source the opportunity to evaluate whether the message produced the intended purpose

Noise: anything that causes a breakdown, interference, or distraction within the communication process; as in electronic communication, noise in interpersonal communication is anything that reduces the fidelity and clarity of the message, such as misunderstanding or misinterpretation by both receivers and senders, preconceived value judgments on the part of both sender and receiver, etc.

---

into the degree of fidelity that the message has. Message fidelity is rarely perfect. Feedback, therefore, gives the source the opportunity to evaluate whether or not the message has achieved its intended purpose.

In the *feedback process,* the receiver and the sender switch roles and the entire communication process begins again. Feedback must not only be complete, but also objective, to be useful to the receiver. Certain active listening and paraphrasing responses, to be discussed in Section 8.7, provide the sender with helpful techniques that assure all concerned that the message has achieved maximum fidelity and that the communication process is complete. Communication is complete when the sender compares the initial concept with the final feedback message.

Effective communication implies that a receiver's understanding of the meaning is equivalent to the sender's intent. Achieving effective communication is not an easy task, nor is it one that is solely the responsibility of the sender if sender and receiver have agreed to be effective with one another.

8.4 BARRIERS TO EFFECTIVE COMMUNICATION

The characteristics of a matrix organization heighten the possibility of communication difficulties. The matrix project team normally consists of a number of individuals from distinct occupational specialties. The specialized language, or jargon, that is characteristic of each, creates barriers when messages from one specialty must be translated by another. Training and loyalties of each profession further result in a set of acquired values that creates an occupational frame of reference. This frame of reference affects the way we view such subjects as cost, deadlines, priorities, and technical performance.

Confused communication can result when reporting relationships are not clearly understood or when channels of communication are fuzzy or confused. Naturally, project teams that suffer from geographic dispersion must make a concerted effort to overcome these structural boundaries by a variety of mediums, such as teleconferencing meetings or colocation, whenever possible. Especially in the matrix project environment, the mixture of group separation and multiple lines of communication tends to produce inadequate and, at times, unreliable communication patterns.

The pressures of time, the demands of what appears at times to be an overwhelming workload, and the changing requirements of a dynamic project life cycle, create a highly stressful environment for most project specialists. Compressed schedules and conditions of uncertainty result more often than not in what team players perceive as "inadequate time" for proper or effective communication.

Misunderstandings naturally arise not only when individuals perceive a lack of time but also when there is a lack of specificity in the communication or the message itself. A memo that ends with "ASAP" can mean "as soon as possible" in a number of different ways: "by Thursday, noon," "within a day," or "a month from now." The point is that there is considerable leeway in the project for misinterpretation and, therefore, a high probability of miscommunication. Project planning and scheduling tools, such as a project summary plan, a detailed work breakdown structure, and a high-lighted critical path, serve as concrete means by which to *verify* each specialist's perceptions and understanding of what is to be done, and when it is expected. They are *not* the substitute, but the vehicle through which face-to-face interpersonal communicaiton occurs!

Another barrier to effective project communication is that words have multiple means, and that the meanings of words are not just within the dictionary definition but within the perception of the receiver. *Connotations* imply that words possess different values: positive, negative, or neutral, and that the real value of a word depends on how the listener perceives it. Words and phrases that an individual uses may not evoke the same images in the next person's mind. A word such as "management," for example, may hold one meaning to a CEO and still another for the head of a labor union.

Table 8.4-1   Nonverbal Communication

- Watch facial color and how it changes as people talk and their feelings come through

- Become aware of expressions that convey tensions, doubt, *mistrust,* inattention, etc.

- Listen for *emotional tone,* another important skill; the tone of voice can convey attitudes that can provide clues as to how to deal with a person in a difficult situation

- Listen for pitch, rate, and subtle variations.

Assuming that everyone knows what you are talking about is usually a poor assumption.

Nonverbal communication is a lot more significant to the success of the communication process than we imagine. Just think, for a moment, of the times people have influenced you simply by the way they looked at you! Psychologist Albert Mehrabian found that when one person talks to another, 7% of the message effect is carried by words, while 93% of the impact reaches the listener–receiver through such nonverbal means as facial expressions that convey trust or distrust, changing facial color, pitch or rate, and subtle variations in time. Once we become aware of the importance of what we and others are *not* saying, more effective and honest communication occurs. Nonverbal communication is, indeed, significant in the listening, influencing, and negotiating processes. Just think, for a moment, of the times people have influenced you simply by the way they looked at you! Table 8.4-1 outlines some important points to remember concerning nonverbal communication.

Corporate or organizational climate consists of the feelings of trust and respect among project participants and employees, the practices related to recognition, the quality of supervision, and quality of work life as a whole. Whether singly or in combination, these variables may enhance or destroy communication for the project and the organization.

As illustrated in Figure 8.4-1, one major obstacle to effective project communication is that *project communication is multidimensional* and must flow upward, downward, and across the organization. Studies further indicate that less than the total of all information is passed along through the chain of command. The extent to which managers and team specialists filter unnecessary information or relay important information affects project performance. Bolyes and Wicker [7], for example, in a study examining the filtering of upward information, found that not only was important information, at times, consciously withheld from supervisors but only 83–85% of total information passed up the chain of command was actually considered

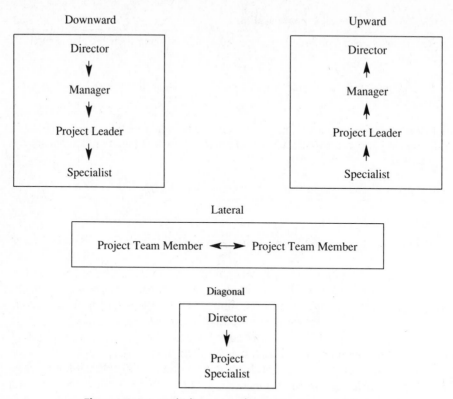

Figure 8.4-1 Multidimensional project communication.

important by specialists and supervisors alike! This infers that in addition to denying Project Managers and supervisors important information, project specialists frequently pass on information that they and their supervisors consider unimportant!

8.4.1 The Relationship between Trust and Accuracy

With the unfortunate events surrounding the disaster of space craft Challenger and its six crew members, a critical question facing Project Managers in the next decade is "Why do project teams find it difficult to present all of the facts?" It is our belief that project teams do not "lie" . . . but fail, at times, to convey relevant information! This distinction between "lying" and not presenting all of the facts is one that has tremendous implications for Project and Functional Managers who desire an open communication environment.

In a previous chapter, we discussed the importance of trust in developing influence and establishing power within a project team or organization. Several studies [8,9] have revealed the degree of trust that exists between project specialists and their various levels of management to be strongly and directly associated with the degree of open, accurate, complete, or undistorted project communication. In addition to trust, the research has explored several other variables, including level of security, autonomy, and overall organizational climate.

More specifically, the literature supported the following assumptions [10]:

- Authority delegated in performing project work is *inversely* related to the distortion of upward communication. That is, where authority is more widely distributed across project team members, *less* distortion is likely to occur.
- The widely held belief that a supportive communication climate helps to build trust was empirically supported [11]. Research results indicate that *accuracy* of communication is *positively* related to organizational climate and *trust*.

The research, therefore, continually supports the belief held by communication theorists and many practicing managers alike, namely, that a supportive climate builds interpersonal trust and that trust leads to more complete and less distorted communication [12].

8.5 ESTABLISHING COMMUNICATION PATTERNS: A SITUATIONAL COMMUNICATIONS APPROACH

Establishing communication networks or patterns is an important procedure in any project setting. Unfortunately, like leadership styles, there is no one best communication pattern. Rather, communication networks are contingent on the nature of the project work involved. They may be developed, however, within such major categories as (1) research projects, (2) technical service projects, and (3) development projects. To manage projects more effectively, managers may need to learn to facilitate communication patterns consistent with the needs or characteristics of their project work.

In a series of studies comparing and contrasting communication networks of high- and low-performing projects, Tushman [13] confirmed that the *characteristics* of the specific project work determine the type of communication patterns and networks that prove efficient.

Research projects, for example, center on work oriented toward developing new knowledge and technologies. Tasks are complex and specialists tend to hold advanced degrees. *Technical service* projects, on the other hand, tend to consider problems specific to the firm's core technologies. Tasks are less complex but tied directly to the firm's source of business. *Development projects,* involving new technologies to deal with a particular product, tend to be closely linked to marketing and manufacturing areas.

Communication patterns for high-performing projects will vary according to the demands of the task and the needs of the specific type of project [14]. High-performing *research projects,* for example, require effective technical feedback and evaluation, and will be closely tied to those areas that provide this kind of input. Communication outside the firm tends to be highly specialized and strongly connected to universities and professional societies. Interface points and roles span the laboratory–professional communication boundaries. "Technological gatekeeping" further serves to keep the project team linked to external information vital to project success. Extensive *decentralized communication linkages* best serve research projects.

Consistent with their task characteristics and professional differences, *technical service projects* are best served by centralized patterns of intra-project communication and greater supervisory input and involvement. Because of the need for input from Marketing and Manufacturing, high-performing technical projects are strongly connected to systems of organizational communication. They are best served when the bulk of information is transmitted through either a supervisory or project focal point. Extraorganizational communication, which focuses on suppliers, vendors and customers, is served through a specific project link.

Consistent with the nature of their task, *development projects* are strongly linked to marketing and manufacturing areas. However, unlike technical service projects, which seem to perform best when information is passed through a specific and direct project focal point, high-performing development projects are characterized by *widespread verbal contact* between specialists within engineering, marketing, and manufacturing. They do, however, require a boundary-spanning project professional (e.g., a Program Manager) to mediate communication with *outside customers, consultants, and suppliers.*

Table 8.5-1 outlines communication networks that prove successful for research, technical service, and development projects.

In developing a model for effective communication processes, Tushman [13] encourages Project Managers and Supervisors alike to consider the three (3) basic sources of "uncertainty":

- The nature of the project task
- The nature of the project environment
- The nature of project task interdependence

"Uncertainty," therefore, depends on the complexity or routiness of the work involved, the rate of change of the environment, and the degree of interdependence with other areas. The greater the uncertainty, the more information networks and processes the project requires.

Table 8.5-1 Communication Networks

	Research Projects	Technical Service	Development Projects
Characteristics and definition	Work-oriented	Consider problems specific to core technologies	Product-oriented
	Complex task		New technologies or extension of existing technologies
	Develop new knowledge and technologies	Tied directly to firm's source of business	Linked closely to Marketing and Manufacturing
	Require advanced degrees		
Communication networks	Closely tied to areas that provide technical feedback and evaluation	Greater supervisory input	Widespread verbal contacts between/among socialists
	Strongly connected to universities and professional societies	Strong link or focal point between Marketing and Manufacturing	Strong link between Marketing and Manufacturing
	Interface points span across the laboratory–professional boundaries	Extraorganizational communication focuses on suppliers, vendors, and customers via centralized project focal point	Strong project professional focal point for extraorganizational communication (i.e., suppliers, vendors, customers).
	Extensive direct, decentralized communication links		

In setting up a communication system, a Project Manager would be wise to:

1. *Analyze information processing requirements for each project, based on the uncertainty posed by the task and its environment.* The greater the complexity and uncertainty of the project task(s), the greater and more frequent will be the communication needs.
2. *Determine appropriate intraproject communication patterns.* Research projects, for example, requiring substantial internal information, are best served by more intense decentralized communication patterns, while projects facing less uncertainty should have less intense and more centralized communication patterns.
3. *Determine the interdependence and degree of differentiation between the project and other areas of the company or agency,* and *choose appropriate internal communication links.* A project may require a formal communication focal point when it consists of areas that have
   • dissimilar time frames, and
   • conflicting goals.
4. *Analyze the project's external communication requirements and choose the most appropriate link.* Research projects require more direct professional communication; development and technical service projects, however, need more central project links.
5. *Periodically review and evaluate the project's information requirements and update as needed.*

Communication patterns certainly have an important impact on project performance. Since Project Managers can influence the communication networks, managers of effective projects will develop communication networks to meet the specific demands of the work involved.

8.6 NEGOTIATING FOR PROJECT SUCCESS

In all project management systems, and especially in those that are of a matrix nature, there is shared responsibility, authority, and accountability. Project participants, often drawn from different operating, administrative, and technical sectors, possess differences in their perceptions. These differences may be related to the techniques or processes to be employed, or the resources, time, and effort that need be expended. Project Managers must often negotiate, not only among specialists and managers contributing to their projects, but also with other Project Managers in settling problems associated with project priorities and responsibilities.

Negotiation is a pervasive process and is an important part not only of business transactions, but of *everyday life*. Negotiating in a project situation may involve such concrete issues as budget, resource allocations, schedules, and/or technical methodology. It may also include intangibles such as recognition, shared product ownership, a sense of accomplishment or achievement, and building continuing effective relationships. One criterion of the successful negotiator is the ability to adapt negotiating style to fit the needs of the specific situation and individuals involved [14]. This requires a firm grasp of the personalities and group dynamics that come into play in each negotiation. Certainly, a knowledge of the negotiating process will prove vital to project success.

Effective negotiators recognize that "winning" with regard to the immediate issue is *not* the totality of the negotiating process. Short- and long-range implications, and the risks associated, must be carefully weighed against the "success" of an instant win. An astute negotiator, for example, may assess that a present "loss" may be a future "gain." Project Managers who operate in a matrix environment must remember that the need to win in the short term may not be as advantageous as maintaining harmonious relationships for the duration of the project.

8.6.1 The Negotiating Process

Negotiation is a two part process. The first part deals with all activities, beginning with the initiation of a request, demand, or offer, and ending when agreement has finally been reached. The latter part of the process concerns the post negotiation period and is, in many instances, considered to be the most important aspect [15]. This is the period when the solution is implemented, when any misunderstandings or hard feelings that still remain may escalate, and when relationships improve or worsen.

Naturally, most negotiators strive to "win" in the first phase of the negotiating process. In fact, negotiating strategies are generally classified around the act of winning. These strategies are labeled according to the results achieved by the negotiating parties. Thus the three basic strategies are: "win–lose," "lose–lose," and "win–win." Table 8.6-1 provides an overview of the characteristics of each of these strategies.

Effective negotiators realize that merely "winning" in the immediate phase of the negotiation process is *not* what the art of negotiation is all about. Short- and long-range considerations must be taken into account. Risks must be calculated and assessed before actions are taken. Although "losses" should be minimized for all parties involved, it is possible that an astute negotiator may want to consider taking an immediate "loss" to support a long-term "win."

Many of us have been indoctrinated with the concept of "winning" through our sports and leisure activities. Golf, tennis, chess, bridge, and most other leisure activities concern a "winner" and a "loser." But negotiators for systems design or product development, for example, must

Table 8.6-1 Characteristics of Negotiating Styles

Win–Lose, Lose–Lose	Win–Win
Controlling orientation (we vs. they)	Problem Orientation (we vs. the problem)
One party's gain is seen as the other party's loss	Mutual gain viewed as attainable
Argument over positions leads to polarization and entrenchment	Seeking various approaches to end, increasing changes for agreement
Each side sees issue only from its own point of view	Parties understand one another's point(s) of view
Short-term approach, focused only on immediate problem	Long-term approach, seeking good ongoing relationship issues
Usually considers only task issues	Considers both task and relationship issues

realize that a ''loss'' may be as advantageous as a ''win,'' if it maintains harmonious and productive relations in the long run.

As indicated by Table 8.6-1 the ''win–win'' strategy naturally has several advantages. Both parties achieve a resolution that makes each feel like a winner. That is, each obtains what they perceive that they need and want from the negotiating situation. This approach is, of course, contingent upon developing cooperative attitudes between all negotiating parties, which is, in and of itself, not an easy feat.

''Win–lose'' and ''lose–lose'' strategies are frequently used in business negotiations. They evolve from our competitive nature, past behaviors, a resistance to change, and a failure to understand how *all parties can win* [16]. In project team efforts they may stem from a lack of understanding of the common goal and a failure to review the effects of our actions on the project organization.

8.6.2 Examining Negotiation Strategies

By understanding the effects of negotiation strategies on attaining project goals, a Project Manager will be better able to choose the strategy that is most appropriate to the situation. The key issue in each of these strategies is *not how much* each negotiator obtains, but *the degree of satisfaction felt as a result of the process.*

Win–Lose. The typical win–lose strategy regards negotiation as a form of competition. In many win–lose instances, the balance of power between or among negotiating parties determines the outcome of the negotiation. Coercion, ignoring the issue, or using some form of majority rule are all examples of win–lose strategies. Voting, although a frequently used win–lose strategy, may prevent team members from becoming personally involved in the solution and may have a negative effect in the aftermath situation. Often, in the win–lose approach, one party's gain is seen as the other party's loss. Frequently, discussions revolve around positions, rather than seeking strategies for mutual gain. This short-term approach tends to focus on the immediate problem and may, at times, prove to be harmful to long-term efforts.

Lose–Lose. In lose–lose strategies the negotiating parties each believe that in order to obtain consensus and achieve a common goal, each must forfeit something they want. Lose–lose strategies frequently are conciliatory in nature. They stem, many times, from a basic desire to avoid conflict and smooth over issues. In this way, compromise and arbitration may, indeed, be regarded as lose–lose strategies; that is, everyone gives up something in the bargaining for the good of the common goal. But again, it is *not* how much one wins or loses that makes a negotiation a success. Success can be measured by the degree of satisfaction each party attains. Naturally, compromise may bring some degree of satisfaction and is, at times, quite necessary in negotiation. However, if project specialists or Functional Managers feel as though their loss was not worth the gain, project commitment will suffer.

Win–Win. Win-win strategies are characterized by a "we versus the problem" orientation. Mutual gain is viewed as attainable, as negotiators strive to generate the various alternative approaches that lead to the common goal. Parties attempt to understand one another's viewpoint(s) and seek a long-term approach, which no one really opposes. The objective of the win–win strategy, actually, is to find a solution that each participant can live with! Win–win strategies are those that seek consensus and begin with an attitude of mutual gain. The win–win strategy is a long-term approach that naturally builds good relations.

Table 8.6-2 describes steps to be taken in win–win negotiation.

8.6.3 The Postnegotiation Period

Satisfaction with a win or loss in negotiating is only part of the total process. The real success or failure occurs *after* the negotiation, when plans or systems need be implemented. Agreements reached, therefore, must be within the acceptable range for all parties involved if negotiation is to be considered *successful*. When contemplating negotiating strategies, balancing the importance of the instant gain to the long-term win is paramount to success.

The way in which the negotiation process occurs, the postures taken,

Table 8.6-2 Five Steps of Win–Win Negotiation

1. Identify the needs of conflicting parties   (Avoid taking positions, i.e., arguing over means, and instead try to identify the ends or needs of both parties. Put yourself in the person's position and ask "why?" as well as "why not?" Recognize the need for respect)

2. Brainstorm a list of possible alternatives–solutions   (Instead of working against one another—"How can I defeat you?"—work together against the problem)

3. Evaluate alternative solutions   (Work for an answer to meet the important needs of all the parties)

4. Implement the solution   (Make sure everyone agrees to and understands it)

5. Follow up on the solution   (Revise, if necessary)

---

words chosen, and emotions displayed are all important pieces affecting this process and our perception of our victory or loss. Implementation of settlements can be enhanced or discouraged by what has transpired during the negotiation. Harmonious relationships need to be maintained and credibility of agreements established. Achieved feelings of satisfaction in *all* parties is the objective of successful negotiation.

8.6.4 Tactics for More Effective Negotiation

Many "tactics" or strategies have been developed to assist in the process of negotiation. These tactics must be chosen wisely to fit the needs of the particular circumstances. In all instances, however, maintaining effective communications and good working relationships is essential to effective negotiation. Some tactics for negotiation are:

1. Whenever possible negotiate on home ground or a neutral site.
2. Prepare an agenda. A well-planned agenda adds structure and purpose to the negotiation and is a benefit to the negotiating parties. Provide the agenda well in advance of the meeting. If supplied with an agenda by another negotiator, submit input or acceptance promptly.
3. Consider the seating arrangement. Arrangement of seating can have a direct effect on the negotiation process. Sit in view of others as well as your team members.
4. Prepare and ask questions to probe feelings; discover attitudes and determine negotiating positions.
5. End on an up beat note! Summarize points of agreement and convey a positive tone.

Some steps to further consider in preparing for a effective negotiation are:

1. Determine what you truly need.
2. Determine what you would like.
3. Establish trade-offs between what you would like and what you really need.
4. Know your negotiating partner! Research (if possible) problems the other participants may have.
5. Record what occurred during the negotiation.
6. Maintain a warm negotiating climate.
7. Practice listening and patience.
8. Focus on the issues.
9. Set out to obtain the whole package, but be prepared to make concessions and trade-offs.
10. Maintain good relations.

Since negotiation is almost a daily occurrence in a matrix project environment, the effective Project Manager must continually maintain organizational cooperation and harmony. In this way, instances calling for negotiation may be made more positive and resolutions may actually improve project relations and performance.

8.7 LISTENING FOR BETTER COMMUNICATION

Listening is one of the most powerful ways to stroke co-workers, customers, or management. Effective listening creates a better environment for negotiation and problem solving. People who are listened to feel valued and important.

Effective listening is as important to the communication process as effective sending. Most "hear" a lot of things, but few of us really listen.

Effective listening is an active give-and-take process. To truly listen effectively, the listener must interact with the speaker in an attempt to understand both content and meaning of the message. This requires a mental and verbal paraphrasing of the message with careful attention given to nonverbal cues. Failure to listen effectively not only prevents adequate responses or counterdemands but also affects the business relations that are established in the process. Let's examine the example below.

Example 8.7-1. Listening Builds Good Relationships and Productivity

One day Ted Burros, a Systems Engineer working in the Naval Sonar Buoy Systems Project, invited Joan Brown from the development group to his area to explain his department's plans for the new Sonar Buoy Antennae System. He described how he thought the system architecture typically used should be changed. Joan's only response was silence and a frown. Ted realized something was wrong and sensed Joan might have something to say.

"Joan," he began, "You have been in the business just about as long as I have. What's your reaction to my suggestion? I'm listening."

Joan paused—eyed Ted mistrustfully—but then began to speak. Ted had opened the door to communication and she felt comfortable offering her suggestions from years of experience. She was also in a more relaxed and trusting mood to be open for Ted's countersuggestions.

As the two exchanged ideas, a mutual respect and trust developed, along with solutions to many of the technical problems.

Several deterrents stand in the way of effective listening. All people have difficulty listening when:

Dealing with a conflict situation.
Dealing with an emotional colleague or customer.
Criticism is being directed at them.
Being disciplined.
Feeling anxious, angry, or even fearful.

Other deterrents to effective listening include: assuming in advance the subject is uninteresting, mentally criticizing the speaker, or overreacting to certain words or phrases.

8.7.1 Active Listening for Better Professional Relations

Communicating and negotiating effectively requires constant *testing and evaluating of assumptions*. Reik [17] suggests that in order to listen actively, the listener should: suspend judgment for a while, develop a sense of purpose and commitment to listening, avoid distractions, develop a *paraphrasing* "active listening" strategy, review the central theme(s) of the message, and continually reflect on what is being said.

"Active" listeners are effective listeners because they become *involved* in the communication process. Listeners, in their own words, restate or paraphrase the message for the sender. In doing so, the sender receives feedback concerning the "fidelity" of the message and if the communication process is, indeed, complete.

Active listening develops a deeper appreciation of what the other person is thinking and feeling. Paraphrasing and asking questions is a good way to help clarify meanings whenever they might be in doubt.

Let's examine the following "Yes . . . But . . ." situation.

"YES . . . BUT . . ."

Gary: "I just don't know what I'm going to do about my boss. He's always picking on me for little things I do wrong."

Kent: "You should talk to him about why your're upset."

Gary: "Yeah, but I couldn't do that. He'd make life miserable for me."

Kent: "Well, you ought to ignore him not let him bother you."

Gary: "Yeah, but then I'd be letting him get away with his lousy behavior and he'd never change."

Kent: "Well . . . you should quit and get another job."

Gary: Yeah, but I need the money, and the way the job market is these days, I probably wouldn't be able to find another job for months."

Kent: (by this time completely exasperated) "Why don't you take a gun and shoot him."

Notice that everytime Kent generated a solution, Gary was spending his listening time thinking of rebuttals and the two co-workers established a circular and frustrating listening pattern!

To be a more effective listener, especially when confronted with a problem or an emotionally upsetting situation, one must take an active responsibility to understand both the *content* of and the *feeling* behind the message.

$$\underbrace{\text{Content} + \text{Feeling}}_{\text{Message}} \xrightarrow{\text{EMPATHY}} \text{Active Listening}$$

Now, let's see what happens when Kent uses a more empathetic "active" listening response.

Gary: "I just don't know what I'm going to do about my boss. He's always picking on me for little things I do wrong."

Kent: "Sounds like you have trouble handling your boss when he points out things you do that he doesn't like."

Gary: "Yeah, and he does it a lot. I don't want to tell him about it because it might make him mad. Then he'd probably make life miserable for me."

Kent: "Hmmm. Boy, that's a double bind. On the one hand you want to tell your boss what you don't like, and on the other hand, you don't want to tell him because he might get upset with you."

Gary: "Yeah, that's exactly how I feel."

Kent: "It's a tough spot to be in. What kind of choices do you have? Let's talk about them."

Notice that Kent's *summarizing* of Gary's response got him out of the "yes

Table 8.7-1   Benefits of Active Listening

Encourages honesty

Develops better understanding

Creates a nonjudgmental atmosphere

Encourages greater information and insight (for you and them)

Defuses some *hot* situations

Creates a feeling of security

Encourages feelings of self-confidence

Reduces stress and tension

Develops mutual respect and trust

. . . but . . ." trap and gave Gary an opportunity to really discuss what was bothering him. Kent responded to Gary's feelings by restating the problem in his own words. As he continued to do so, he moved the interaction toward a problem-solving plan.

Empathetic Active Listening is an excellent technique to defuse a conflict situation, understand a negotiating opponent, or better generate customer wants, needs, and requirements. The active listening process accomplishes several things for both "them" and you.

They:

- Get if off their chests.
- Feel better understood.
- Might gain more insight.
- Solve their own problem.
- Are in a better frame of mind to accept any advice.

You:

- Know them better!
- Can be better informed to give more appropriate advice.

But active listening requires *genuine* concern for the individual as a person. It also means you must be neutral on all subjects, and take an objective

Table 8.7-2   Active Listening Techniques for Better Professional Relations

Nonverbal listening responses	Eye contact, facial expressions, gestures (e.g., nod), body position
Casual remarks	"I see," "uh-huh," "go on," "say more about that," "tell me more"
Echo	Repeating verbatim the last few words of the speaker: "okay, from what you say, I understand . . ."
Questioning	Encouraging further contribution: "what was your reaction to that?" or "when did this begin to occur?"
Reflecting	Letting the speaker know you are empathetic (e.g., you feel angry, concerned, frightened, happy, etc.) "you seem concerned about the schedule and the changes that have to be made"
Paraphrasing	Stating in your own way what the message means to you: "so you think the order will be delayed because of these changes"

stance. It requires more time to listen and to relate, and means you *never* make the decision for them, but help to verbalize their decisions!

Table 8.7-1 outlines some benefits of active listening. Table 8.7-2 lists some suggestions for active listening techniques for better professional relations.

8.8 CONCLUSION: IMPROVING PROJECT COMMUNICATION

Project Managers must recognize that the environment that they create within the project structure has an effect on the successful communication of information. Communication is enhanced by clearly defined channels, well-established reporting mechanisms, and good working relationships. Communication breakdowns can be avoided if individuals know with whom they should communicate and what information needs to be transmitted.

Communication breakdowns often result from a variety of problems. These include:

- Confused lines of authority.
- Frequently changing organizational structure.
- Lack of proper planning.
- Top-heavy, cumbersome hierarchies.

Communication:
Top Project Priority
Consider:

What to Say?
To Whom?
When?
How?

- Wide spans of control.
- Lack of proper communication skills in project specialists.
- Physical distance.

If the Project Manager truly wishes that the project team communicate more freely and frequently, a trusting environment must be secured that encourages upward, diagonal, and lateral communication. The more important and respected project specialists feel, the greater will be their willingness to communicate.

To encourage more effective communication, Project Managers must practice and model two-way communication strategies, demonstrate a sincere interest in what the specialists have to say, and reward project teams for their efforts to communicate. Requesting reports and information and providing feedback in a regular and timely manner further improves the communication process.

REFERENCES

1. Kezsbom, D. S. "Communications on Communications (C2)," *Communications Society Magazine, Vol. 20, No. 3, May 1982, pp. 47–48.*
2. Kezsbom, D. S., "Oral Communication and the Listening Process," *Communications Society Magazine,* Vol. 20, No. 4, July 1982, p. 37.

3. Tushman, M. L. "Managing Communication Networks in R&D Laboratories," *Sloan Management Review,* Vol. 20, Winter 1979, pp. 37–49.

4. Leslie, C. E. "Negotiating in Matrix Management Systems," in D. Cleland (Ed.), *Matrix Management Systems Handbook.* New York: Van Nostrand, Reinhold, 1984.

5. Rosenbloom, R. W., and Walek, F. W. "Information and Organization: Information Transfer in Industrial R&D," Boston, MA: Graduate School of Business Administration, Harvard University, 1967.

6. Sayles, L. *Managerial Behavior,* New York: McGraw Hill, 1964.

7. Boyles, D., and Wicker, H. "Deliberate Filtering of Upward Communication," in Barndt S., "The Matrix Manager and Effective Communication," in D. Cleland (Ed.), *Matrix Management Systems Handbook.* New York: Van Nostrand, Reinhold, 1984.

8. Might, R. "An Evaluation of the Effectiveness of Project Control Systems," *IEEE Transactions on Engineering Management,* Vol. EM-31, No. 3, August 1984.

9. Barndt, S., "The Matrix Manager and Effective Communication," in D. Cleland (Ed.), *Matrix Management Systems Handbook,* New York: Van Nostrand Reinhold, 1984.

10. Athanassades, J. "The Distortion of Upward Communication in Hierarchical Organizations," *Academy of Management Journal,* Vol. 16, 1973, pp. 207–226.

11. Muchinsky, P. M. "The Interrelationships of Organizational Communication Climate," as cited in Barndt, Ref. 7, p. 584.

12. Barndt, Ref. 7, p. 384.

13. Tushman, M. L. "Managing Communication Networks in R&D Laboratories," *Sloan Management Review,* Vol. 20, Winter 1979, pp. 37–49.

14. Leslie, C. E. "Negotiating in Matrix Management Systems," in D. Cleland (Ed.), *Matrix Management Systems Handbook.* New York: Van Nostrand, Reinhold, 1984.

15. Ibid. p. 631.

16. Kirchof, W. S., and Adams, J. R. *Conflict Management for Project Managers.* Project Management Institute, PA, 1986.

17. Reik, T. *Listening With The Third Ear.* New York: Pyramid, 1972.

EXERCISES

Nonverbal Communication

Take a moment to reflect on the following questions and write your responses.

How do people let you know nonverbally that they aren't listening?

How do they let you know that they have a problem?

How do they let you know that they want to terminate the conversation?

How do they let you know that they aren't interested in what is being said?

How do they let you know that they have responded to "red flag" words?

How do they let you know that they are daydreaming?

Activity: Barriers to Effective Project Communication

1. Outline the process of

Upward   :
Downward :
Lateral   :   Communication in your project organization.
Diagonal :

2. Determine both tangible and intangible project outputs (memos, status reports, problems, pitfalls, etc.) that relate to each specific form of communication.

3. What are the *barriers* that "block" or prevent effective communication at this level (barriers may include fears, intimidation, lack of procedures, lack of information, etc.).

4. Generate activities and solutions that may overcome these barriers and create a more productive project environment.

Hint: Do you see an interesting *trend* in the solutions? Improved interpersonal communication improves interorganizational communication as well. If we effect a change in our upward communication mechanisms, we effect a similar change in downward communication.

Listening Practice Exercise

The following situations will give you a chance to practice identifying empathetic active listening responses. Circle the responses in each situation that you think are emphatic. Check your answers at the end of the exercise.

1. "It happens every time the manager appears in my department. He just takes over as if I weren't there. When he sees something he doesn't like, he tells the employee what to do and how to do it. The employees get confused and I get upset.

a. "You should discuss your problems with your boss."

b. "When did this start to happen?"

c. "The boss must be the boss, I suppose, and we all have to learn to live with it."

d. "It can be upsetting when one manager overrides another's authority. It's difficult to confront it, too."

2. "It's happened again! I was describing an office problem to my boss when she started staring out the window. She doesn't seem to be really listening to me because she has to ask me to repeat things. I feel she's superficially giving me the time to state my problems, but she ends up sidestepping the issue."

a. "You should stop talking when you feel she's not listening to you. That way, she'll start paying attention to you."

b. "You can't expect her to listen to every problem you have. Anyway, you should learn to solve your own problems."

c. "It's frustrating to have your boss behave this way, especially when you talk about problems that are important to you."

d. "What kind of problems do you talk to her about?"

3. "I think I'm doing all right, but I don't know where I stand. I'm not sure what my manager expects of me, and he doesn't tell me how I'm doing. I'm trying my best, but I sure wish I knew where I stood."

a. "Has your boss ever given you any indication of what he thinks of your work?"

b. "Not knowing if you are satisfying your boss can really make you feel insecure on the job."

c. "Perhaps others are also in the same position. Don't let it bother you."

d. "If I were you, I'd discuss it with him."

4. "He used to be one of the guys until he was promoted. Now he's not my friend anymore. I don't mind being told about my mistakes, but he doesn't have to do it in front of my coworkers. Whenever I get a chance, he's going to get his!"

a. "To be ridiculed in front of others is always embarrassing."

b. "If you don't make so many mistakes, your boss wouldn't have to tell you about them."

c. "Why don't you talk it over with a few people who knew him before and then go talk to him about this situation?"

d. "How often does he do this?"

Answers: **1.** *d* **2.** *c* **3.** *b* **4.** *a*

Practice Sessions—Empathetic Listening

Using the empathetic listening approach, design listening statements in response to the following situations. Compare your response with those at the end of the exercise.

Situation 1. John, a fellow supervisor who is conscientious and strong-willed, has lost the ambition to be productive. He responds irritably to routine problems and as result has become difficult to work with. You are talking to him during lunch when he says to you: "Nobody cares. There are no efforts being made to improve conditions. We get no information or leadership from management and we seem to have the same problems over and over."

Your response: _____

_____

_____

_____

Situation 2. You are a supervisor in the Corporate Quality and Manufacturing Department. One of your engineers, Derek, is outspoken, fault finding, and demanding. His coworkers respond to him by being offended and avoiding him as much as possible. Derek says to you: "The people who work with me are lazy and unfriendly. I tell them what they ought to do, but they don't listen to me."

Your response: _____

_____

_____

_____

Situation 3. A fellow supervisor, who is conscientious and responsible, says to you in an angry voice: "I don't know how to deal with this program manager. He doesn't seem to listen to me when I tell him I'm not able to meet the date for his requested action. I have to wait for the vendor to deliver the parts and yet he won't accept my reasons why the action cannot be accomplished."

Your response: _____

_____

_____

_____

Discussion of Practice Exercise—Empathetic Listening

Here are some possible empathetic responses for the situations given above.

Situation 1. "Seems you feel discouraged about the way things are being handled, and that you're fed up with the same old problems."

Situation 2. "If I understand you correctly, Derek, your coworkers don't do things the way you feel they should. How do you let them know what you think they ought to do?"

Situation 3. "It's annoying to think you've explained your problem in a way the other person should understand, but he won't seem to listen to how you are restricted."

CHAPTER 9

MANAGING TEAM PERFORMANCE [1]

9.1 INTRODUCTION

Those of us who are sports fans most likely have a favorite team. If we're avid fans, we attend at least an occasional game, watch them on TV, or listen to the radio sportscast. When the game is over, who better than we to provide a detailed account of the team's victory or defeat. What irritates us, and most avid fans, is to watch a team *fail* to play together. We can see immediately when, for example, one misses a block, fails to pass on information to the quarterback, or tries to shine at the other team members' expense. We are experts in diagnosing the team's weak spots and in providing feedback to the coach! With all our insight into teamwork and team development, many "project coaches" fail to see the parallel between what is needed to improve, for example, a football team and what is needed to shape up their own project teams!

Effectively managing and influencing the diversity of personnel responsible for outlining and implementing project components clearly is a key to a Project Manager's success. In fact, more than in any other form of organizational work, project teams because of their multidisciplinary and interdependent nature demand highly attuned team building skills by which they can achieve project objectives. Considering the national fervor for team sports, it is surprising to find the low level of carryover into organizational life. Project teams have much to learn from the playing field.

9.2 WHAT IS A TEAM?

Let's start out with a few basic concepts. The most common response that we encounter in our team building seminars to the question "What is a team?'' is ''A group or people, with a common goal.'' Such a common and *erroneous* assumption of what a team is prevents us from truly developing a group into a productive and cohesive team. A team is much more than merely a group of people that perceive a common goal. A team is a collection of people who must rely on cooperative group effort and on the specific skills and abilities of each interdependent team player. All groups are not teams, but unfortunately all teams can be merely a group!

A team is a special designation awarded to a group of people who are aware of the very nature of their interdependent roles and how the skills each possess complement their efforts and assure goal attainment. Effective team output is best explained through the concept of synergy. Through effective teamwork, a group can generate solutions to problems that are far superior to those developed individually by its team members.

9.2.1 Essentials of a Team

Moving from a ''group'' to a ''team'' is a distinction achieved only through awareness and effort. Managers, like coaches, must be able to examine team effort with a skilled eye and determine what is blocking maximum output. Then, with the help of their ''players,'' they must devise a strategy to overcome or remove the obstacles and combine team resources to achieve project goals.

Teamwork is not a mystical process. From the research on group dynamics and the process of team building, we know a great deal about how teams operate. In addition, we can pinpoint the essential elements that lead to successful team performance, which are basically a ''charter,'' a ''mission,'' a reason for working together; a sense of interdependence; commitment to the benefits of group problem-solving and group decision-making; and accountability as a functioning unit.

For any group to function effectively as a team, several important characteristics must be present. First, the group must have a ''charter'' or a reason for working together. The members of the group must *need* each other's experience, abilities, and commitment in order to achieve success. Group members must be committed to the idea that working together as a team leads to more effective decisions than does working in isolation.

Participation in planning and decision-making help team members to develop a sense of ownership and commitment to team goals and procedures. Each team member must be held in regard for the special talents and abilities that that person brings to the team and is encouraged to contribute to team strategy sessions. Team players, with the help of the ''coach,'' recognize their respective roles and functions within the framework of the team and realize that through these roles, project goals *are* attainable. Communication is characterized by candor, and feedback is directed at specific team-related actions and not at personalities. Listening occurs for understanding rather than defense. Finally, the group is accountable as a functioning unit within the larger organization and, as such, demonstrates pride in their accomplishments.

9.2.2 Characteristics Of Effective versus Ineffective Teams

Why are some teams so successful while others appear to be less coordinated or effective? It's sad but true that all of us could list at least a dozen or so characteristics of ineffective teams as a result of either direct personal experiences or close contact with a losing team. Ineffective teams often waste much of their energy defending themselves against feedback and inappropriately competing with each other. Players tend to demean or diminish one another rather than collaborate toward achieving common goals. Instead of supporting and encouraging new ideas, ineffective teams dwell on personality factors that are neither relevant to the task or just cannot be modified. Energies that should be directed toward "winning" are funneled into wasteful practices that lead to negative output.

Effective successful teams, however, are characterized by clear, well-understood priorities and plan for ample opportunity to discuss and clarify team objectives. Project team members individually must understand their priorities in relationship to project objectives, and *how* to accomplish them. Leadership is consistent and appropriate not only to the project and team player needs, but to the skills required by the team as it proceeds through its process of development.

Integrating personal and team goals is a major task of the effective team leader. For the team as a whole to accomplish its goal, team members must be able to achieve their personal goals within the larger framework of team objectives. Motivation is at its maximum, as team members see opportunities to grow and to develop.

It is true that the quality of the team's output is as good, if not better than, the synergy of the diversity of ideas it generates. The proper blend of the variety of skills and perspectives represented enables the project team to achieve solutions to diverse problems by drawing on the right person at the right time. While such diversity usually leads to conflict (or disagreements), the successful resolution of the disagreements creates stronger solutions and an even more effective project team.

Productive teams are aware that the climate, or quality of team interactions, affects not only morale but also productivity and performance. A climate that produces respect, innovation, and excellence fosters a winning team. Successful teams not only attend to detail, anticipate problems, and follow through on plans, but continually strengthen the bond of friendship and respect celebrating their accomplishments.

Creating a well-blended, effective team does not require the elimination of team "stars." Top players, however, learn not to dominate the game, but use subtle techniques to accomplish their objectives. A well-blended team is like a well-running piece of machinery; each part of the machine must interact smoothly with the next for the machine to operate. If one part is out of synchronization, the machine will cease to function. The real "secret" of team blending is knowing how to make the assist, when to pass to someone in a better position than you are, and what will improve not only your position, but that of your team partners.

9.3 THE PROCESS OF TEAMWORK—STAGES OF TEAM DEVELOPMENT

Teams tend to go through a natural process of development that a skilled or trained eye can easily identify. Team building activities should be sequenced and planned according to this natural order.

Stage I—Establishing Identity. This phase of team development, often referred to as "forming" or "inclusion" [2], is characterized by team member's attempts to find their place within the group. To assist the group through this phase of team development, the Project Manager, or team leader, should:

- Allow ample opportunity for team members to get to know each other.
- Affirm and legitimize the distinctive abilities and strengths of each participant.
- Clarify work expectations and rules that will govern team performance.
- Agree on the major mission and determine the objectives and priorities of the team.

Failure to perform these tasks may cause team members to fixate or get stuck in this phase and never progress to maximum team performance.

Stage II—Questioning Authority. This phase is often characterized by the "storming", "struggling" and "in-fighting" that surfaces as like-minded individuals form factions and coalitions that question team leadership and control. Project leaders in this phase are best served by listening and responding calmly and fairly to member challenges while mediating between factions and allocating the workload according to member skills and task requirements. Mediation becomes the primary goal of the project leader. If the split between team factions is not properly bound, team development will never quite reach full maturity.

Stage III—Productivity. Once team members feel relatively comfortable with each other, and mechanisms for task accomplishment and resolving differences are in place, the team can focus on the primary objectives of *quality and quantity*. Now, project participants understand precisely what

Table 9.3-1 Stages of Team Development

Stages	Characteristics	Leadership Skills and Behavior
I. Establishing identity	Team members attempt to find their place within the group	Allowing ample opportunity for team members to get to know each other
		Affirming and legitimizing the distinctive abilities and strengths of each participant
		Clarifying work expectations and rules that will govern team performance
		Agreeing on the major mission and determining the objectives and priorities of the team
II. Questioning authority	Struggles and in-fighting surface as like-minded individuals form factions and coalitions	Mediating between factions
		Listening to and responding calmly and fairly to member challenges
	Coalitions question team leadership and control	Allocating work according to member skills and task requirements
III. Productivity	Team focuses on its primary objectives	Facilitating resolution of problems
		Injecting fun and variety into the work setting
IV. Uniting	Successful integration of team efforts	Providing opportunities for team to celebrate accomplishments
	Comradeship	
	Team members celebrate accomplishments	

they must do and set out to accomplish it. Leadership becomes much more *participative,* as the leader's task involves facilitating the resolution of problems and of injecting fun and variety into the work setting. Failure to demonstrate appropriate leadership strategies may result in boredom or complaints concerning non-work-related issues.

Stage IV—Uniting. Feelings of attraction and comradeship grow from the successful integration of team efforts. Team members enjoy their association with one another and come together often to celebrate their accomplishments. Providing opportunities for the team to celebrate the fruits of their labor is the most important leadership requirement of this final phase of team development.

Table 9.3-1 provides an overview of the characteristics and skills associated with the four basic stages of team development.

9.4 THE TEAM DEVELOPMENT PROCESS

Team building is the process of creating and then maintaining effective team functioning. Team development takes time and commitment. The results, however, are great in terms of higher morale and performance, increased productivity, and more innovative problem-solving.

Team development is concerned not only with the cost, schedule, and technical performance parameters but also with the human interactions and feelings that arise during the project effort. A major objective of team building is to assist teams in managing their task *and* interpersonal concerns.

Team development begins with the group's recognition that it is dealing with significant issues related to improving team effectiveness. Norms supporting candor, openness, and trust are reinforced early on through successful experiences in participative problem-solving and decision-making. Providing opportunities for the development of trust facilitates problem-solving by increasing the exchange of relevant information and open discussion.

One fundamental ingredient of team development is full participation of each team player in accomplishing whatever objective the group sets out to achieve [3]. The most effective means of implementing any plan, strategy, or procedure is by encouraging the full participation of those who will be responsible for its final implementation. Participation translates into commitment and creates a psychological bond between the plan or strategy and those who generate it. This *is* the "buy-in" that Project Managers, especially those working within the matrix project organization, strive to accomplish. If commitment is the desired outcome, then participation in the broadest sense must be encouraged.

Not just the Project Manager, but *all* participants must be sensitive to the participation level of each team player. If participation is low or nonexistent, team leaders must actively solicit and encourage full participation. Such sensitivity to the team's needs will ensure a high level of participation and all the benefits it brings.

9.4.1 Steps in the Formal Team Building Process

The formal team building process consists of activities carried out in the early stages of the project and then integrated into *ongoing* project activities. Because there is usually little time to set aside for special team building procedures, team building must often be incorporated into planning sessions, review meetings, and informal updates or discussions on project status.

The following are steps for a formal team development process [4].

> *Step 1—Establishing a Positive Environment.* Setting a positive tone or climate for a team building process involves helping participants to understand team building and generating commitment to the benefits of the process. It involves defining the goals of team building, determining how team building can contribute to project objectives, and determining what it requires in terms of time and commitment. Many team members, especially those in an engineering or technically oriented environment, may erroneously believe that team building is a "touchy–feely" process and is of no consequential value to the *real* issues of the project. If such is the case, care must be taken to explain that the objective of team building is *to increase the team's productivity and performance*. These objectives are accomplished by:
>   - focusing on the team mission
>   - determining key tasks and responsibilities, and
>   - developing team roles.
>
> *Step 2—Developing a Sense of Interdependence.* A critical step in the initiation of a team building process is to develop a sense of interdependence and respect for team players' complementary talents. Providing opportunities to discuss team players' backgrounds and illustrating how these diverse backgrounds contribute to the fulfillment of the team mission will help to accomplish this important team building step. As early as the project kick-off session, Project Managers should *provide opportunities for team players to understand and feel confident in the abilities of their comembers.*
>
> *Step 3—Define and Clarify Team Goals.* It is critical to the project's success that team members understand and accept project goals. To establish goal orientation, it is important to *jointly develop a mission statement defining what the team expects to accomplish.* Specific objectives concerning costs, schedule, and performance criteria are generated from the mission. To increase the likelihood of accom-

plishing project goals, objectives should be reviewed, discussed, and clarified at team building meetings. As each objective is discussed, the team strives to deal with disagreements and synthesize agreements. Should conflict occur, plans must be developed concerning strategies for dealing with the specific conflict issues. Goals must be clearly understood by each team player. They should, in fact, be as concrete (or measurable) as possible and applied directly to a time line or schedule.

Step 4—Role Definition. If roles remain fuzzy, not only frustration, but duplication of effort, and conflict may result. Each function's responsibility must be clearly determined, including any informal agreements made between team players. Responsibilities must be defined in terms of accomplishing the project mission, objectives and subobjectives. Major expectations regarding mutual roles are established and communicated and difficulties between perceptions discussed and clarified. New roles, if necessary, are negotiated and developed. Team members concentrate on the strategies that each team contributor, (e.g., Marketing, Engineering, Manufacturing) can use to best accomplish their respective roles.

Step 5—Developing Procedures. A team cannot possibly reach its full effectiveness if procedures are not thought out. Guidelines and policies must be developed for recurring as well as special issues. By establishing proper guidelines and procedures, the team helps to minimize conflict and concentrate on obtaining project objectives. Who should attend review meetings, how cost data should be tracked, or when status reviews should be conducted are just some of the many procedures that must be determined as early as the project kick-off meeting.

Step 6—Developing a Decision-Making Process. This final step in the team building process involves determining responsibilities for decision-making; that is, determining *who* should be directly involved in the decision-making process and *how* they will be involved. This will, of course, be affected by the scope and importance of the decisions.

One technique to accomplish these six important steps that the authors have applied and refer to as *process management strategies* (PMS). Process management strategies are directed toward developing clearer understanding of the team's mission and determining *collectively* what the team is *paid* to do. It involves determining how team roles (Marketing, Engineering, Manufacturing, etc.,) contribute to the project mission or to accomplishing project objectives and is led by objective, experienced facilitators.

Through process management strategy sessions, key managers involved in implementing project objectives are responsible (with the assistance of an outside, objective facilitator) for determining and agreeing on what is to be done, how it is to be done, and who is to do it. Mission statements generated by the management team define the boundaries of the project and identify the critical elements or things that *must* be done in order to accomplish the mission. Through brainstorming sessions, team members explore both internal and external opportunities and the threats (or risks) in accomplishing their mission. Consensus is reached on the critical factors leading to the success of the team and the strategies to be taken to assure these factors are achieved [5].

9.4.2 Developing Team Roles

Without thinking, many team members play habitual roles; that is, they behave in ways most comfortable for them on the basis of their past experience in groups. This is done regardless of the real needs of the team and without deliberate forethought. To be effective team players, members must be able to recognize the various *task* and *maintenance* behaviors that are *required* for smooth and productive team performance. They must further be able to fulfill these roles and avoid negative or dysfunctional behaviors [6].

Some important team roles and common behaviors associated with them are:

Task Roles	Behaviors
Initiating	Proposes tasks or goals Defines group problems Suggests ideas for solving problems
Seeking information and opinions	Requests facts Seeks information about group problems Solicits opinions Asks for suggestions or ideas Asks for expression of feelings
Giving information And opinions	Provides relevant facts and information about group problems States belief about a matter
Elaborating	Gives examples Develops meanings Indicates how a proposal might work out
Summarizing and orienting	Restates suggestions Pulls together related ideas Offers decision or conclusion for the group to accept or reject
Evaluating	Analyzes ideas and determines feasibility Evaluates accomplishments of the group

Maintenance Roles	Behaviors
Encouraging	Is friendly and responsive to others Accepts others and their contributions
Expressing (feelings)	Expresses feelings present in the group Calls attention of the group to its reactions about ideas and suggestions
Gatekeeping	Keeps communication channels open Facilitates participation of all members Suggests procedures for discussing group problems
Setting standards	Expresses standards or goals for the group to achieve Helps the group to become aware of its direction and progress
Compromising	When own idea is involved in a conflict, offers compromise Admits error Disciplines self to maintain group cohesion
Harmonizing	Attempts to reconcile disagreements Gets people to explore their differences Reduces tension through joking, relaxing comments

Dysfunctional Roles	Behaviors
Dominating	Demands submission from others Does not ask what others wish or think Tries to take over the discussion
Withdrawing and/or avoiding	Does not participate in group processes Withholds ideas and suggestions Refuses to express own feelings when needed
Blocking	Rejects the view of others Overdoes "devil's advocate" role Kills new ideas through negative responses Uses emotion-laden language and encourages disagreements
Recognition seeking	Seeks personal recognition for contributing to group effort Tries to monopolize discussion Does not note group conditions Special interest pleading

Dysfunctional Roles	**Behaviors**
Topic-jumping	Continually changes the subject Makes irrelevant comments Moves ahead without checking for integration of ideas

In dealing with dysfunctional roles as, for example, the "blocker" or the "dominator," the Project Manager as well as other team members will find it helpful to follow three general steps:

1. Draw attention to the *behavior* itself. Avoid labeling or classifying the person as, for example, a "sniper" or a "handclapper." Such labels will only elicit defensiveness. Instead, describe the behavior that's getting in the way.

   Example: For one who dominated unduly the intentions of a group, it is better to say "You talked a great deal in the meeting" rather that "You're a dominating loudmouth!"

2. Spell out, in a *supportive* manner, specific dysfunctional *effects* of the behavior especially on team performance. Often, one who distracts the group is unaware of the negative impact of his or her behavior. Sometimes that person really *wants* to be making a contribution and does not know how to be an effective team member.

   Example: "When you dominate the conversation, I feel that you prevent others from expressing their ideas. We need all the suggestions we can get in designing this product."

3. Finally, suggest exploring *alternative behaviors* that lead to a more productive and satisfying climate.

   Example: "What are some alternative courses of action you can take to make your suggestions known?"

Remember when dealing with dysfunctional roles, focus your constructive feedback on the *behavior itself* and *not* the "personality" of the individual. Encourage the change process by helping the individual seek alternative behaviors that produce more satisfying interactions.

9.5 THE ROLE OF THE PROJECT MANAGER

For the team development effort to be successful, Project and Functional Managers must view the process as an opportunity for increasing their own effectiveness and status. Team members, in addition, must also understand the importance of a team development effort, desire the effort, and somewhat understand the process.

As previously mentioned, team development is not a single event. It may begin with a kick-off session meeting or a formal team development workshop, but it must be carried on throughout the life of the project and seen as the means toward productivity and improved quality of work life. Moving the team from immaturity to maturity is the primary goal of a team building effort.

A critical part of the manager's job is to plan a strategy that will be effective in reversing any negative outputs. The first step is to critically review the team's performance by asking some basic questions. If you're afraid to ask, you already know the answers!

Organizational Outputs	**Yes**	**No**
1. Are your profit margins, sales, quality ratios, and production quotas down?	—	—
2. In the R&D environment, are new suggestions, products, or development declining?	—	—
3. In service units, has the level of customer utilization declined below the desired level?	—	—

Organizational Processes	**Yes**	**No**
4. Are employees subject to constant criticisms, isolation, conflict, or overcontrol?	—	—
5. Have grievances or complaints among the staff increased?	—	—
6. Is there evidence of hostility among the team members?	—	—
7. Are assignments frequently misunderstood or not carried out properly?	—	—
8. Is there a feeling of apathy and general lack of interest or involvement among staff members?	—	—
9. Are there constant one-on-one dealings with the boss and little staff interaction?	—	—

The original team development strategies were created primarily to facilitate the resolution of issues centered mostly on matters of conflict, disruption, ineffective methods, unclear assignments, or failed expectations of leadership. These are generally areas where something needs to be repaired or some problem solved. However, sometimes the issue is not a matter of an unsolved problem or that of disruption, but a case where people have drifted into a routine, noninnovative, standard way of doing business. Creating a winning team involves a continual attempt at maintaining effective team relations and molding a group of strangers into a workable unit. It is one of the most exciting and difficult challenges a Project or Functional Manager will face.

Some management strategies for developing this sense of team spirit include:

- Providing accurate and continual feedback to the group about its performance.
- Including team members in the goal setting and decision-making process.
- Keeping channels of communication open among team players, perhaps with regular meetings or informal discussions.
- Encouraging supportive communication.
- Developing mutual understanding of roles and responsibilities.

9.6 SELECTING TEAM MEMBERS

Prior to actual project personnel selection, it is necessary to first define key functions and the resources required to accomplish these functions. Work definitions can be generated from the work breakdown structure (WBS). On the basis of the scope of project work, the Project Manager must determine personnel needed in the Project Office and must prepare job descriptions for these functions. It must further be decided whether these positions should be allocated as full-time to project efforts or are shared among other projects or functions.

Functional support personnel must be obtained, in most cases, through negotiations with functional management. In order to successfully accomplish these negotiations, the Project Manager must first define the tasks to be accomplished, the level of expertise needed to accomplish these tasks, and when and how long the individuals will be needed. Project planning inputs, such as the Statement of Work (SOW), the WBS, and, of course, the schedules are all important to this effort.

Reporting relations must be defined with input and agreement from contributing functional managers. These procedures should define to whom the specialists is formally accountable, the standards of performance, schedule, budgets, and the salary and performance review process. This understanding that has been established between Project and Functional Managers must be communicated to the appropriate project personnel and adhered to throughout the project.

The project team must be task organized to meet project objectives. This implies that one of the Project Manager's primary job functions is to identify what tasks need be done, what functions of the organization may be affected by the project, and what skills are needed for the project to be brought to successful completion.

Effective Project Managers should be aware of some common team member selection problems. Problems frequently encountered include:

- Not delineating all functions or skills needed to successfully accomplish project objectives.

- Late entry into the project.
- Premature entry of a team specialist.

Of course, these problems add to delay, cause confusion, and increase costs.
   Good working teams should be lean enough to be a hardworking, tight-knit team yet broad enough in skills to accomplish project objectives. There should be, moreover, flexibility to add or drop members based on project requirements.

Table 9.7.1   The Project Kick-Off Meeting Agenda

December 28, 1987

Location:–West Wing Conference Room B

1.0	Introduction and kick-off comments by senior management	8:00–8:30
2.0	Introduction of team specialists	8:30–9:15
3.0	Project overview, mission, and procedures	9:15–9:45
	Coffee Break	
4.0	Project organization	10:00–10:30
5.0	Discussion of project goals and major milestones	10:30–11:15
6.0	Establishing individual commitments (building a responsibility chart)	11:15–12:00
	Lunch	
7.0	Review project summary plan	1:00–2:45
	Review customer inputs and requirements	
	Review first two levels of the WBS	
	Review functional plans (e.g., R&D, design, reliability and test, manufacturing, etc.)	
	Break	
8.0	Discuss criteria for success and team rewards	3:00–4:15
9.0	Schedule subsequent status review meetings	4:15–4:30

9.7 ORGANIZING TEAM MEMBERS

Organizing the project team involves implementing the project plan by assigning specific responsibilities to specialty functions or work packages to team members. The Responsibility Matrix, which was discussed in Chapter 2, helps to pinpoint, publicize, and clarify team member responsibilities, as it is circulated among the various team players.

The project kick-off meeting is the Project Manager's first opportunity to initiate the team building process. It is exactly at this point that the Project Manager must attempt to involve team members and generate commitment to a common goal. The objectives of the project kick-off meeting are to:

1. Introduce project team members.
2. Define the team mission and develop project goals.
3. Determine reporting relationships.
4. Define lines of communication.
5. Establish or review project plans.
6. Pinpoint high-risk or problem areas.
7. Delineate responsibilities.
8. Generate and obtain commitment.

Because the project kick-off meeting is the Project Manager's first attempt at team building, it is *extremely* important that attendees leave the meeting with a sense of organization, stability, and accomplishment. A carefully planned agenda, like that illustrated in Table 9.7-1, helps to assure the Project Manager and meeting participants that these objectives will be accomplished.

9.8 IDENTIFYING AND OVERCOMING BARRIERS TO TEAM PERFORMANCE

Identifying barriers that impede effective team performance is a major requirement for building a more effective project team. Some common team performance barriers[9] and probable causes include:

Barriers	Causes
Differing outlooks, priorities, and interest	Different professional interest
Role conflicts	Ambiguity regarding assignments; lack of planning
Unclear objectives	Lack of mission, understanding of market, or clearly defined customer requirements

Barriers	Causes
Dynamic project environment	Characteristics of projects themselves; changing project priorities due to customer requests, top management demands, regulatory changes
Competition regarding team leadership	Lack of team definition and structure; lack of management and leadership skills in the Project Manager; technical problems or problems in overall project effort
Lack of team definition and structure	Poorly defined tasks, responsibilities, and reporting relationships; several contributing organizational elements without a project focal
Team personnel selection	Unavailable or limited resources that produce inappropriate assignments, low motivation, and uncommitted members
Credibility of project leader(s)	Inadequate managerial skills, inconsistent and confusing actions, poor communication or technical judgments, lack of experience or inadequate track record
Lack of team member commitment	Poor match between professional interests and project objectives; inadequate or inappropriate reward structure; insecurity associated with project effort; poor relations between Functional and Project Managers
Communication problems	Multiple causes—difficult to determine; may include insufficient time, carelessness, lack of commitment, lack of procedures, or poor relations
Lack of senior management support	Senior management may fail to understand project environment and/or project needs; lack of timely feedback; shifting priorities

9.8.1 Overcoming Barriers to Team Performance

Successful Project Managers are not only aware of potential problems that may arise in the project environment but also are constantly on the lookout

Table 9.8-1   Barriers to Team Performance: Causes and Recommended Actions[a]

Barrier	Causes	Recommended Actions
Differing outlooks, priorities, interests of team	Multidisciplined team; different professional interest; multiproject environment	Provide overview of project scope through WBS and project plan; attempt to match project objectives and personal goals of project specialists; sell team concept; define roles and responsibilities early
Role conflicts	Ambiguity over responsibilities; unclear, multidiscipline matrix environment	Provide opportunity for team members to submit input to WBS; assign and negotiate roles early on in project efforts (e.g., kick-off session); reinforce roles through linear responsibility charts and through frequent status review meetings
Unclear project objectives	Lack of mission; unspecified market or customer; poor communication among Marketing, Engineering, and senior management	Jointly establish team mission; define team objectives; develop a common sense of purpose perhaps through team name and logo; communicate frequently with top management and other contributing project organizations
Dynamic project environment	Changing requirements due to customer demands or senior management needs; regulatory changes; characteristics of life cycle	Stabilize dynamic influences; forecast environment, perform risk analysis; develop contingency plans; establish a Change Control Board and invite user–customer participation
Competition regarding team leadership	Lack of team definition and structure; weak leadership skills in Project Manager;	Clear role and responsibility definition; support of Project Manager by senior management; formal

Table 9.8-1 (*Continued*)

Barrier	Causes	Recommended Actions
	problems in project effort and technology	training for Project Manager in leadership, influence, etc.
Lack of team structure and definition	Poorly defined roles and responsibilities; unclear reporting relationships; no project focal point	Education in team essentials and team building process; regular team meetings; linear responsibility chart; provide visibility and recognition
Project personnel selection	Limited resources or lack of talent; matrix environment; turf issues	Early negotiation of roles early on with Functional Managers and team specialists; present overview of project in kick-off session; replace those who are really not interested; job posting; alignment of project formally within organization
Credibility of project leader	Lack of managerial or technical skills or experience; wavering or inconsistent values; conflict between Project Manager and functional departments; previous track record	Develop good working relationships with key project contributors; train and develop personnel; choose from informal leaders, and provide with appropriate skills and authority.
Lack of team member commitment	Poor match between team player abilities and project needs; insecurity regrading project effort; inadequate reward structure	Role model enthusiasm by Project Manager; match skills with needs of project; determine reasons for project insecurity; resolve conflicts that may exist between team members; team rewards

Table 9.8-1 (*Continued*)

Barrier	Causes	Recommended Actions
Communication problems	Logistics of matrix organization; geographic dispersion; failure to determine responsibilities and reporting relationships; insufficient time; poor interpersonal relations and carelessness; personality clashes	Teleconferencing; timely meetings; communication tools, reporting mechanisms; colocation; frequent communication with contributing functions and with customer; social activities; clearly defined and delegated work packages
Lack of senior management support	Lack of understanding of project management needs; shifting priorities; unclear strategic mission; budget limitations	Invite senior management to kick-off sessions; frequent newsletters; keep management informed of milestones and successes

[a] Adapted from Kerzner, H., and Thamhain, H. *Project Management for Small and Medium Sized Businesses*. New York: Van Nostrand Rheinhold, 1984.

for symptoms and conditions that encourage such deficiencies. Effective Project Managers, therefore, are constantly working toward building a work environment conducive to team building. Project Managers who not only possess managerial skills to effectively plan, organize, direct, and administer project efforts but are sensitive to the barriers that impede team performance have a clear advantage towards successful project completion.

Table 9.8-1 provides an overview of some common causes of poor team performance, [7] and recommended actions to overcome them.

9.9 CONCLUSION

Effective team performance comes from creating an environment in which team members work together under a unity of purpose, as a united front. A team building effort should be directed at:

- Establishing a better understanding of the team's characteristics—its purpose and role in the total functioning of the organization.
- Creating a better understanding of each team member's role in the work group.

- Increasing communication and creating greater support among team members.
- Discovering more effective ways of working through problems at both task and interpersonal levels.
- Developing the ability to use conflict in a constructive rather than destructive manner.
- Encouraging greater collaboration among team members and the reduction of costly competition.
- Creating a sense of interdependence.

With the growth of the project matrix organization, high-technology corporations of the future will require that a person belong to several units at the same time and have two, three, or even four bosses. People will shift rapidly from one project team to the next. If such is the case, the organization with managers who know how to build effective teams quickly, will have that special advantage. Team effort will continue as long as humans must rely on others to achieve results.

REFERENCES

1. Kezsbom, D. S. "Creating A Winning Team," *IEEE Communications Magazine,* Vol. 21, No. 6, September 1983, pp. 50–51.
2. Berger, M. "The Technical Approach to Teamwork." *Training and Development Journal,* Vol. 39, No. 3, March 1985, pp. 53–55.
3. Hall, H. "Systems Maintenance: Gatekeeping and the Involvement Process," in J. A. Shotogren (Ed.), *Models for Management: The Structure of Competence.* The Woodlands, TX: Telemetric International, 1980, pp. 285–301.
4. Wilemon, D. "Developing High Performance Matrix Team," in D. Cleland (Ed.), *The Matrix Management Systems Handbook.* New York: Van Nostrand Reinhold, 1984, pp. 363–378.
5. Hardaker, M. and Ward, B. K. "Getting Things Done. How to Make a Team Work." *Harvard Business Review,* Vol. 65, No. 6, November–December 1987, pp. 112–119.
6. Glaser, R., and Glaser, C. "Building a Winning Management Team," Bryn Mawr, PA: Organization Design and Development, Inc., 1980, 1986.
7. Thamhain, H., and Wilemon D. "Team Building in Project Management," *Project Management Quarterly,* June 1983.

Exercise 1:
Structured Experience

Members

Roles

TASK ROLES													
Initiating													
Information Seeking													
Information Giving													
Elaborating													
Orienting–Summarizing													
Evaluating													
MAINTENANCE ROLES													
Encouraging													
Expressing Feelings													
Gatekeeping													
Standard-Setting													
Compromising													
Harmonizing													
ANTIGROUP ROLES													
Dominating													
Withdrawing–Avoiding													
Blocking													
Recogniton-Seeking													
Topic-Jumping													

Structured Experience: Team Member Role Nominations Form

Exercises

EXERCISE 1: STRUCTURED EXPERIENCE: TEAM MEMBER ROLE NOMINATIONS

After your next project team task, attempt a simple, brief (1-hour), and useful exercise to help team members see what must occur if a rewarding job is to be accomplished. Through this exercise the concept of team roles and functions is illustrated and members may come to see that better team behavior can be easily in reach.

After the group performs a team task, the Project Manager or team facilitator provides "role nomination forms" to each team contributor and has each member mark the column(s) corresponding to the roles they perceive their colleagues play most often. All members should include themselves. The group then appoints an individual to record and post the team data. The team then proceeds to discuss the varying behaviors demonstrated and make generalizations towards productivity and task performance. Teams may discuss, for example, who supplied what functions, whether all functions on the sheet were filled, the relative flexibility of members, and so on.

The worksheet form for this structured experience on team member role nominations can be found on the preceding page.

Instructions: For each member, place check marks in the column corresponding to the roles that person has played most often in the group. Include yourself.

EXERCISE 2: TEAM BUILDING EFFECTIVENESS SURVEY

Getting a handle on team performance begins by assessing where the team perceives it is now and where it should be in the near future.

Prior to a formal team building session, the Project Manager may seek information from team players on their perception of team functions. These functions include goals, roles, procedures, relationships, and outside environmental influences. The information obtained through the completion of *anonymous* surveys is collated and used as the basis for team discussions and action plans.

Team-Building: A Group Effectiveness Questionnaire

Instructions: This survey is designed to identify how you feel your team has been functioning. Specifically, what areas you feel need improvement. As you read the issues, think about how your group has been working and respond as openly and honestly as possible.

Where there are multiple choices, check only one answer. Please be as candid and specific as possible.

Goals

1. List the goals that require members of the team to work together.

   _____
   _____
   _____
   _____

2. With regard to the goals of this team, the people on this team are:
   _____Very committed _____Somewhat committed
   _____Somewhat resistant _____Very resistant
   Team goals that I feel have low commitment are:

   _____
   _____

   Team goals that I feel have high commitment are:

   _____
   _____

3. To what extent do you know and understand each team member's goals?
   _____Very knowledgeable about team member's goals
   _____Fairly knowledgeable about team member's goals
   _____Somewhat vague knowledge of team member's goals
   _____Very vague knowledge of team member's goals
   Goals I would like to know more about are:

   _____
   _____

4. The goals of members of this team are:
   _____Strongly in conflict _____Somewhat in conflict
   _____A little in conflict _____Not at all in conflict
   The key conflicts I see are:

   _____
   _____
   _____

Roles

1. The roles of the members of the team are:
   _____Very clear to me _____Fairly clear to me
   _____Somewhat unclear to me _____Very unclear to me
   Areas I would like to have clarified concerning my role are:

   _____
   _____

   Areas in which I am unclear about what others expect of me are:

   _____
   _____
   _____
   _____

   Areas I would like to have clarified concerning others' roles are:

   _____
   _____
   _____
   _____

2. Team members' roles overlap:
   __Very much __Quite a bit __Somewhat __ Not at all
   Roles that overlap are:

   _____
   _____
   _____

3. Team members' roles that are in conflict are:

   _____
   _____
   _____
   _____

Procedures

1. The decision-making process on matters that affect more than one member of the team is:
   _____Very clear _____ Fairly clear
   _____Somewhat unclear _____Very unclear
   Decisions that need to be clarified are:

   _____
   _____

2. Thinking of all the communications within this team, I would say they are:

_____Good with all members of the team
_____Good with some members of the team
_____Good with very few members of the team
_____Not very good with the majority of the team
Please specify what subjects you feel need to be better communicated:

_____
_____
_____
_____

3. Team meetings
    _____ Are very effective
    _____ Are fairly effective
    _____ Need some improvement
    _____ Need much improvement
    Please specify how you would like team meetings to be improved.

    _____
    _____
    _____

4. I believe the leadership of this team
    a. Is helping the team's performance by:

    _____
    _____
    _____

    b. Could improve the team's performance by:

    _____
    _____
    _____

5. As teams work together, behavior patterns are established that help or hinder the team's performance such as:
    Following up or not following up on decisions
    Raising or not raising sticky issues
    Facilitating or delaying decisions
    Do you feel that there are patterns that inhibit this team's effectiveness?
    _____Yes _____No
    If yes, what are they?

    _____
    _____
    _____
    _____

Relationships

1. Conflicts within the team are:
   _____Openly discussed and resolved
   _____Discussed somewhat
   _____Very seldom mentioned
   _____Not discussed at all
   A conflict that needs to be resolved to improve the team's performance is:

   _____

   _____

   _____

   _____

2. Relationships I would like to discuss are:

   _____

   _____

   _____

   _____

3. What is causing stress for you and/or other members of the team?

   _____

   _____

   _____

   _____

Environmental Influences

1. What constraints or influences outside of the team keep it from working more effectively? Please explain.

   _____

   _____

   _____

   _____

General

1. What do you believe are the team's key strengths?

   _____

   _____

   _____

   _____

2. If you could, what would you do to make this team more effective?

   _____

   _____

   _____

   _____

CHAPTER 10

HOW TO WRITE A WINNING PROPOSAL

10.1 INTRODUCTION

This chapter is intended for the engineering group responding to a Request-For-Proposal (RFP) from either a government agency, another company, or another division of your company. In addition, much of this chapter can be used by the engineer in responding to upper management's request that a given specification be met within a particular time and for a prescribed cost; however, in such a case some of the details presented below can be omitted.

10.2 THE PREPROPOSAL PHASE

10.2.1 The Request For Proposal (RFP)

A U.S. government-sponsored RFP's availability is always announced in the *Commerce Business Daily* (CBD). Interested companies can write to the appropriate procurement person and receive a copy of any RFP thought to be of interest. Usually the RFP must be responded to, by a proposal, within 30 days. Requests for proposals are also sent by most large corporations, and government organizations to companies that are on their bidder's list.

If the CBD announcement or a copy of the RFP sent to you from a bidder's list is the first time that you noticed the project to be performed, then your chance of winning the contract is very small! The purpose of good marketing is to seek out the development of programs of prospective customers such as the government or other companies. Once Marketing becomes aware of a program development effort, which often occurs up to 2 years prior to the RFP, Engineering and Marketing should meet regularly with the prospective customer in an attempt to better understand the customer's need and to help formulate the specifications and statement of work. Such assistance is warranted and appreciated, since if enough prospective contractors review the specifications and discuss the problems with the customer, the customer, too, becomes more knowledgeable about the problems, the project duration, and the possible cost of the project. Indeed, very often the government issues a "request for comments on specifications" in order to obtain official information from possible contractors concerning how to best perform a particular project.

If the marketing arm of your company does its "scouting" properly, the Engineering group will completely understand the specifications and the statement of work having influenced the content of these documents. Further, initial steps in writing the proposal, such as setting up the proposal writing team, performing a work breakdown structure (WBS), as well as some initial design work, can be completed prior to the issuance of the RFP.

10.2.2 To Bid or Not to Bid

To bid or not to bid is, indeed, the question. Writing the proposal costs money. A rule-of-thumb is that a good proposal costs approximately 5% of the eventual contract award.

The cost of writing a proposal is an indirect charge to the company as a *General and Administrative* (G&A) *Expense*. It is clear that losing too large a percentage of bids will severely increase one's G&A, making it more difficult to bid the next RFP at a competitive cost. Further, winning a bid requires that personnel, facilities, and other company resources be allocated to the resulting contract. If such an allocation could be used on a different contract to produce a greater income, a business decision must be made as to whether the RFP should be responded to.

To facilitate the required business decision, an in-depth look at the RFP by Marketing and Engineering is required. This technical information is then provided to the management responsible for the *bid–no-bid* decisions.

10.2.3 The Sole-Source Proposal–White Paper

When Engineering or Marketing believes that it has a good idea for a new product, that idea should be investigated. If a single customer is being considered for the product, that customer should be sent a "white paper" that briefly describes the idea and then be visited and the concept discussed. The prospective customer, Marketing, and Engineering, can put together the required specifications and discuss a reasonable timeframe for delivery and the approximate price of the item. The white paper is then expanded on and becomes a *sole-source* proposal, that is, a proposal submitted by only one contractor. The proposal format used is similar to the technical and cost proposals discussed in Sections 10.5 and 10.7.

If the idea is for a product to be marketed nationally and/or internationally, it is then necessary that Engineering and Marketing "sell" their idea to upper management. If the "idea" appears promising, in a business sense, Marketing will be requested to perform a market analysis and Engineering a feasibility study. This material is further augmented as described in Sections 10.5 and 10.7 to form the "white paper proposal."

10.2.4 Small Business

A small business is one that has sales less than a prescribed amount and/or has fewer than a specified number of employees. Although the actual numbers vary, one can usually assume that a small business is one that has annual sales of less than $10 million and/or fewer than 500 employees. If such numbers seem "large" to you, you are, indeed, a small business! Remember, however, that such a definition excludes companies such as AT&T, ITT, TRW, Hughes, Rockwell, and a host of others.

If you are a small business, visit your friendly small-business office that is located at each government facility. The small-business representative will help you contact other small businesses and the customer in an attempt to persuade the customer to restrict the forthcoming RFP to small businesses *only*. There are also small-business offices at many major companies that are intended to give work to appropriate small businesses.

There are other "restrictive" categories in addition to small business. One is a "woman-owned" business in which at least 51% of the ownership is held by women who exert a major role in the company's operation. A second is a "minority" or a "disadvantaged" business in which majority ownership is by minorities or disadvantaged individuals. In this case, the small-business office will be happy to introduce an official of the business to the customer, and if the business is qualified, the RFP can be restricted.

10.2.5 Teaming

If Marketing and Engineering are well prepared, they will know who their competitors are. Remember, these competitors are also "lining up" to see the customer. You are not the only one. Very often, in large and even in small contracts, it is advisable to team with another company having strengths complementary to yours. By broadening the team's strengths, weaknesses are reduced and a better proposal can be generated.

Contractually, the teaming agreement is done by a *subcontract* or a *joint venture*.

10.2.6 Subcontracting and Joint Ventures

A subcontract formalizes the agreement between the team members, spelling out in detail the role each will have when the contract is won. It also specifies how to handle proprietary material that each member may have contributed. Further, it allows the customer to know who the team members are and their individual responsibility.

Very often, a small business responding to an RFP restricted to small business will team up with a large company. In such cases, the small business is required to do a significant amount of the contract work. There is some current discussion concerning the requirement that at least 50% of the work be performed by the small business and this is often a good rule to follow.

A *joint venture* is where the team members form a new corporation for the sole purpose of performing the contractual work. Joint ventures are very complicated legally and politically, since the employees of the venture come from the respective teaming companies and will return to their companies, with the knowledge gained from their performance on the contract, after the contract ends. After the venture dissolves, both companies will often produce the product formed by the venture and compete for the business.

10.3 THE REQUEST FOR PROPOSAL (RFP)

The RFP contains:

- The statement of work.
- The specifications (including quality) to be met.
- A listing of all deliverable items such as hardware, software and reports.
- Instructions on what the technical, management, and cost proposals should contain (some RFPs require only a Technical and Cost proposal, the Management proposal becoming a part of the Technical proposal).
- The rules and regulations that must be adhered to.
- Corporate representations and certifications.
- A point of contact, which is the procurement officer responsible for the RFP.

Once the RFP is available, one often cannot talk to the technical customer. If there are technical questions, they can be sent, in writing, to procurement who will forward all questions to the technical group. After these questions are answered, they are sent, along with their answers, to all prospective bidders, thereby becoming common knowledge. It is therefore often a good strategy to avoid asking questions and just see what questions your competition asks. In this way, you learn what your competitions' weaknesses are.

If a company believes that its proposal is of great value to the government, but the due date is too soon to permit an adequate presentation, a delay can be requested. It is improbable that such a delay will be granted, unless it can clearly be shown to be in the best interest of the government. However, if an amendment (change) is made by the customer, such as requesting that extra work be performed or changing a specification, then it can be sometimes successfully argued that the proposal due date be extended. In general, one should assume that the proposal due date cannot be changed.

10.4 THE PROPOSAL-WRITING TEAM

The proposal-writing team must put together the technical, management, and cost proposals and therefore should consist of a group having diverse talents. If possible, the prospective Project Manager should be selected to be the proposal team leader.

The technical proposal (see Section 10.5) includes a discussion of the conceptual design, the WBS, and all the other required schedules discussed in Chapters 3 and 4. Three to five technical individuals having the primary body of knowledge required to accomplish the contract should be responsible for the proposal preparation. Other specialists should be brought in as needed to give advice on such matters as design, development or manufacturing intervals, and lead times on purchasing, and to perform simulations required for the proposal, and help in the design of the product.

The Management Proposal (see Section 10.6) includes management techniques and controls, as well as an organization plan for the project. Thus, it is recommended that an experienced Project Manager or Corporate Manager write this section.

The Cost Proposal (see Section 10.7) takes the schedules from the Technical Proposal and expands them to include the monthly or quarterly cost of each deliverable, cost of personnel, and other variables. Thus, a person with an accounting background is very useful to the Cost Proposal effort.

Also note that since each of these individual proposals are tied together closely, the Project Manager is the logical person to be the team's "leader."

10.5 THE TECHNICAL PROPOSAL

The Technical Proposal is the most important part of the proposal. It indicates an understanding of the problem, describes several possible approaches, and discusses in detail the proposed approach. In addition, the Technical Proposal includes the program plan, including schedules, personnel, personnel hours, and the cost of materials. In general, the Technical Proposal *must not* include salaries and/or wages of personnel or their cost to the project. Further, each section of this proposal must, wherever applicable, reference the section of the Statement of Work (SOW) being addressed.

A typical table of contents is given in Table 10.5-1. Here the first three items are self-explanatory.

The "Paragraph Cross-Reference" page indicates which sections of the

Table 10.5-1   Typical Table of Contents of Technical Proposal

Table of Contents

List of Figures

List of Tables

Paragraph Cross Reference

1.0 Understanding the Requirements

2.0 Engineering Approach

    2.1 Candidate Approaches

    2.2 Proposed Approach

3.0 Feasibility Approach

4.0 Testing

5.0 Facilities and Personnel

6.0 Work Breakdown Structure and Schedules

7.0 Matrix of Engineering Labor

8.0 Materials

9.0 Subcontracts

Appendix

technical proposal address each section of the SOW and the specifications.

"Understanding the Requirements" is a section written to demonstrate to the customer that you understand the SOW, the specifications, and how the product is to be used. After reading this section, customers should have the feeling that you understand their problems as well as, or better than, they do.

The "Engineering Approach" section consists of two subsections: "Candidate Approaches" and the "Proposed Approach." The *candidate approaches* include all of the different ways that one could possibly solve the problem. It does not, however, include the proposed solution! Each candidate approach is discussed in detail, its pros and cons explained, and the risk of such an approach—in terms of performance, schedule, and time—discussed. In particular, if one is aware that a competitor is bidding a specified approach, it is advisable to include this approach in your list and to explain in detail the drawbacks of such an approach. The "Proposed Approach" explains, in great depth, your approach to solving the problem. This is where the conceptual design, software, and hardware simulations as well as a detailed block diagram (to chip level if possible) should be presented. Remember, here is your chance to tell the customer that you thoroughly understand the problem and how to solve it. Above all, write this section clearly! If it is not understood by the customer, you will have lost your chance to gain the customer's confidence. If you can't make yourself understood now, why should the customer believe that you will be understood later?!

The "Feasibility of Approach" is a study of the risk associated with each element of your design. Most customers want a *low-risk* approach. If there is any uncertainty or if state-of-the-art work is necessary, you must prove to the customer that you can succeed. It is a good idea to include in this section related work, a list of publications, texts, and patents, all of which help to establish your expertise and credibility.

"Testing" is a very important aspect of the technical proposal. It represents the customer's assurance that when the project is completed, the product will operate according to the specifications i.e., that quality will be assured. To convince the customer of your ability and willingness to perform proper testing, you must first explain what you will test for and then how the test will be conducted. These two items form a test plan. Equipment used for testing should be described, and evidence of your having performed similar testing on some other project should be presented.

Now that you have convinced the customer that you are technically capable of solving the problem, you must next demonstrate that you have the necessary facilities and personnel to perform the project work. Describe the facilities' location and the equipment available. Enumerate the personnel to be used in the project. List both full-time employees and consultants and include an abbreviated biographical sketch of each. Further, a proposed project organization chart should be included along with a description of how it "fits" within the corporate structure. Inclusion of a personnel—experience matrix is useful. A typical matrix is shown in Table 10.5-2.

This next section, "Work Breakdown Structure and Schedules," is extremely important. This indicates to the customer that the technical program has been well thought out. The WBS shows each of the activities that must be performed. A Gantt chart showing the duration of these activities and their expected start and stop times is presented as well as a milestone chart showing when each deliverable is due. Further, a matrix of labor versus hours is presented. This matrix consists of three tables, which are shown in Table 10.5-3. The first table presents worker-hours per activity.

Table 10.5-2   Personnel—Experience Matrix

Personnel – Experience matrix

			SKILLS		
Employee	Education	Years of Experience	Physical Design	Software Development	...
Tom Case	BS	5		✓	
John Doe	PhD	15	✓		
Bill Roe	BS	10		✓	
⋮					

Each activity listed is taken from the WBS. To illustrate the use of this table note that the employee, W. L. Second, is expected to spend 600 hours on Activity D and 1720 hours on the project. Also note that Activity D is expected to require 4700 hours for completion. The second table lists personnel hours expended for each quarter, while the third table presents activity hours per quarter. If individual participants are not yet known, the personnel can be listed in terms of labor categories such as Principal Engineer, Research Engineer, Analyst, Computer Scientist, or Technician. If such categories are used, descriptions of these categories should be presented under "Facilities and Personnel."

If "Materials" must be purchased and paid for by this contract, these should be listed with their cost.

If there is to be a "Subcontract" with another company, a statement of work to be performed by the subcontractor must be listed. In addition, the Subcontractor's facilities, personnel, project organization, and so on, must be presented in the appropriate places in the Technical Proposal.

10.6 THE MANAGEMENT PROPOSAL

The Management Proposal is not given as much weight as the Technical Proposal; however, many a prospective contractor has lost a contract because of a poor-quality Management Proposal.

The purpose of the Management Proposal is to:

- Demonstrate to the customer that you have won other related contracts and have performed properly, that is, have met the specifications on time and at cost.

Table 10.5-3 Matrix of Labor, Hours and Activities

Personnel – Hours/Activity

| Personnel | Activity (from WBS) | | | | | |
	A	B	C	D	E	TOTAL
W.M. Head (PM)	120	200	400	600	800	2120
W.L. Second	120	400	200	600	400	1720
P.Q. Third	300	200	400	500	600	2000
A.J. Fourth	200	800	1000	2000	150	4150
C.Q. Fifth	200	800	800	1000	200	3000
TOTAL	940	2400	2800	4700	2150	12,990

Hours/Task

Personnel – Hours/Quarter

| Personnel | Quarter | | | | | | | | TOTAL |
	1	2	3	4	5	6	7	8	
W.M. Head	265	265	265	265	265	265	265	265	2120
W.L. Second	160	160	160	160	240	280	320	240	1720
P.Q. Third	300	240	240	280	240	280	280	140	2000
A.J. Fourth	520	520	520	520	520	520	520	510	4150
C.Q. Fifth	320	320	400	400	400	320	320	520	3000
TOTAL	1565	1505	1585	1625	1665	1665	1705	1675	12,990

Hours/Quarter

Activity – Hours/Quarter

| Activity | Quarter | | | | | | | | TOTAL |
	1	2	3	4	5	6	7	8	
A	940								940
B	625	1505	270						2400
C			1315	1485					2800
D				140	1665	1665	1230		4700
E							475	1675	2150
TOTAL	1565	1505	1585	1625	1665	1665	1705	1675	12,990

Hours/Quarter

- Explain to the customer the management techniques you employ, to control a project and to be responsive to the needs of the customer.
- Describe how your company intends to organize for this particular project.

Specifically the Management Proposal is divided into the following chapters:

Table 10.6-1 Table of Past Contracts

No.	Purchasing Activity and Address	Contracting Officer's Name and Address	Contract No.	Type of Contract	Award Prices/Cost	Final Price/Cost	Period of Performance
1	US Navy, Bldg 1 Crystal City, Va.	Michael Persone Rm 706	USN 86-2567-AP	CPFF	$2,876,500	$3,001,000	6/86 – 6/88
2							
3							
4							
5							
6							
.							
.							
.							

"Table of Contents." This includes a list of tables, figures, and chapter titles.

"Chapter 1—Past Performance."—This chapter is divided into two parts. The first is a tabular listing of all contracts for related and/or similar work. Table 10.6-1 shows a typical table. The second part of this chapter is composed of a brief technical summary of the SOW, complexity, objectives achieved, and an explanation for any cost overruns or delays.

"Chapter 2—Corporate Management Techniques and Controls." This chapter contains a corporate organization chart and a discussion of the management techniques and controls used. For example, is a matrix organization employed? How is it set up? Are there established reviews or meetings? What type? How often? What types of reports are required? Also, the firm's *Earned Value Analysis* (EVA) cost-control technique must be discussed in this section! The government, or any prospective customer who will pay you for work to be done, wants to know how you manage to get your projects completed on time and within cost. Your tracking, controlling, and reporting techniques must be described. Typical blank report or monitoring forms should be included.

"Chapter 3—Project Structure." In this chapter, contractors must explain how they intend to

Organize for the performance of this contract.

Interface the various elements of the organization during performance.

Recognize and react to problems.

Use the EVA cost-reporting system to monitor actual progress and cost and compare them to planned progress and cost.

Control performance and cost.

10.7 THE COST PROPOSAL

The Cost Proposal is intended to indicate to the customer how the cost of the contract is divided:

- Among direct and indirect expenses.
- Between personnel or labor category by month and by task.
- Between tasks by months.
- Cost of materials.
- Cost of reports.
- Cost of travel and subsistence.

Table 10.7-1   Table of Contents

List of Figures

List of Tables

1. Standard Form 1411

2. Summary of Direct and Indirect Costs

3. Matrix of Personnel Cost and Hours by (Month) Quarter

4. Matrix of Personnel Cost and Hours by Task

5. Matrix of Task Cost and Hours by (Month) Quarter

6. Cost of Materials

7. Cost of Reports

8. Cost of Travel and Subsistence

9. Overhead, General & Aministrative Expense and Fee

   Justification

10. Subcontract

A typical Table of Contents is shown in Table 10.7-1. Each topic in the Table is described below:

1. *SF 1411 Form.* This form is shown in Table 10.7-2. Each line shown in SF 1411 is described below:
   a. The solicitation number given in the RFP.
   b. The contractor's name and address.
   c. Usually an individual from the contractor's procurement or contract division and telephone number.
   d. Type of contract action, that is, whether it is a new contract or an engineering change proposal, and so forth.
   e. Whether the contract specifies a Firm Fixed Price (FFP), Cost Plus Fixed Fee (CPFF), Fixed Price with Incentive (FPI), and so on. The FFP contract is usually bid very *conservatively,* since the

Table 10.7.2   SF 1411

CONTRACT PRICING PROPOSAL COVER SHEET	1. SOLICITATION/CONTRACT/MODIFICATION NO.	FORM APPROVED OMB NO. 3090-0116

NOTE: This form is used in contract actions if submission of cost or pricing data is required. *(See FAR 15.804-6(b))*

2. NAME AND ADDRESS OF OFFEROR *(Include ZIP Code)*

3A. NAME AND TITLE OF OFFEROR'S POINT OF CONTACT	3B. TELEPHONE NO.

4. TYPE OF CONTRACT ACTION *(Check)*

A. NEW CONTRACT	D. LETTER CONTRACT
B. CHANGE ORDER	E. UNPRICED ORDER
C. PRICE REVISION/REDETERMINATION	F. OTHER *(Specify)*

5. TYPE OF CONTRACT *(Check)*
☐ FFP   ☐ CPFF   ☐ CPIF   ☐ CPAF
☐ FPI   ☐ OTHER *(Specify)*

6. PROPOSED COST *(A+B=C)*

A. COST	B. PROFIT/FEE	C. TOTAL
$	$	$

7. PLACE(S) AND PERIOD(S) OF PERFORMANCE

8. List and reference the identification, quantity and total price proposed for each contract line item. A line item cost breakdown supporting this recap is required unless otherwise specified by the Contracting Officer. *(Continue on reverse, and then on plain paper, if necessary. Use same headings.)*

A. LINE ITEM NO.	B. IDENTIFICATION	C. QUANTITY	D. TOTAL PRICE	E. REF.

9. PROVIDE NAME, ADDRESS, AND TELEPHONE NUMBER FOR THE FOLLOWING *(If available)*

A. CONTRACT ADMINISTRATION OFFICE	B. AUDIT OFFICE

10. WILL YOU REQUIRE THE USE OF ANY GOVERNMENT PROPERTY IN THE PERFORMANCE OF THIS WORK? *(If "Yes," identify)*
☐ YES ☐ NO

11A. DO YOU REQUIRE GOVERNMENT CONTRACT FINANCING TO PERFORM THIS PROPOSED CONTRACT? *(If "Yes," complete Item 11B)*
☐ YES ☐ NO

11B. TYPE OF FINANCING *(√ one)*
☐ ADVANCE PAYMENTS ☐ PROGRESS PAYMENTS
☐ GUARANTEED LOANS

12. HAVE YOU BEEN AWARDED ANY CONTRACTS OR SUBCONTRACTS FOR THE SAME OR SIMILAR ITEMS WITHIN THE PAST 3 YEARS? *(If "Yes," identify item(s), customer(s) and contract number(s))*
☐ YES ☐ NO

13. IS THIS PROPOSAL CONSISTENT WITH YOUR ESTABLISHED ESTIMATING AND ACCOUNTING PRACTICES AND PROCEDURES AND FAR PART 31 COST PRINCIPLES? *(If "No," explain)*
☐ YES ☐ NO

14. COST ACCOUNTING STANDARDS BOARD (CASB) DATA *(Public Law 91-379 as amended and FAR PART 30)*

A. WILL THIS CONTRACT ACTION BE SUBJECT TO CASB REGULATIONS? *(If "No," explain in proposal)*
☐ YES ☐ NO

B. HAVE YOU SUBMITTED A CASB DISCLOSURE STATEMENT (CASB DS-1 or 2)? *(If "Yes," specify in proposal the office to which submitted and if determined to be adequate)*
☐ YES ☐ NO

C. HAVE YOU BEEN NOTIFIED THAT YOU ARE OR MAY BE IN NON-COMPLIANCE WITH YOUR DISCLOSURE STATEMENT OR COST ACCOUNTING STANDARDS? *(If "Yes," explain in proposal)*
☐ YES ☐ NO

D. IS ANY ASPECT OF THIS PROPOSAL INCONSISTENT WITH YOUR DISCLOSED PRACTICES OR APPLICABLE COST ACCOUNTING STANDARDS? *(If "Yes," explain in proposal)*
☐ YES ☐ NO

This proposal is submitted in response to the RFP contract, modification, etc. in Item 1 and reflects our best estimates and/or actual costs as of this date.

15. NAME AND TITLE *(Type)*	16. NAME OF FIRM
17. SIGNATURE	18. DATE OF SUBMISSION

NSN 7540-01-142-9845

1411-101

STANDARD FORM 1411 (10-83)
Prescribed by GSA
FAR (48 CFR) 53.215-2(c)

Attachment 3

U.S. GOVERNMENT PRINTING OFFICE : 1984 O - 437-443

customer and contractor have agreed to delivery at a specified price. The CPFF contract is often bid *optimistically* since the contractor agrees to perform at his best effort. If, however, unexpected problems are encountered, the customer agrees to accept the work produced for the contract price or provide extra funding to complete the project. This extra funding does not include profit (fee) since the fee is *fixed*. FPI is a fixed-price contract in which the customer agrees to share the additional funding required should cost overruns occur. The share contributed by the customer decreases as the overrun increases.

f. The proposed cost consists of (i) the direct and indirect costs, (ii) the profit per fee (typically 7–13%), and (iii) the sum (i + ii).

g. Contractor's address and contract duration.

h. The line item numbers are the deliverables. Each deliverable item should be identified, and its total price presented. The sum of all the items in 8D should be the same as the total given in 6C.

i. The address of Defense Contract Administration Services Management Agency (DCASMA) and the address of Defense Contract Audit Agency (DCAA).

j–m. These are self-explanatory;

n. The Civilian Acquisition System Board (CASB) data is available from the Federal Acquisition Regulations (FARs), Part 30.

o–r. These are self-explanatory.

2. *Summary of Direct and Indirect Costs.* A typical summary is shown in Table 10.7-3.

3. *Matrix of Personnel Cost and Hours by Month or Quarter.* This is an extension of the matrix presented in the Technical Proposal. An illustration of the matrix is given in Table 10.7-4.

4. *Matrix of Personnel Cost and Hours by Task.* This is illustrated in Table 10.7-5.

5. *Matrix of Task Cost and Hours by Quarter.* This is illustrated in Table 10.7-6.

6. *Cost of Materials.* All materials purchased for and charged to the contract must be listed here with their price.

7. *Cost of Reports.* The deliverables in a contract include documentation, a major category (usually line 2) in the WBS. In this section, each report is listed together with its estimated pages, number of figures, cost of typing and reproduction, etc., and the cost set forth. Thus, the output of this section is the cost of reports. This section does *not* include engineering personnel if it is included in the Personnel Cost sections.

8. *Cost of Travel and Subsistence.* Each contract will usually require travel. Thus, each trip and the number of people per day must be

Table 10.7-3   Summary of Direct and Indirect Costs

(1)	Direct Costs	$10,100,000
(2)	Overhead (Burden) (<u>100</u> % of(1))	10,100,000
(3)	Other Direct Costs	
	Consultants	1,000,000
	Equipment	2,000,000
	Travel	100,000
	Subcontracts	1,000,000
(4)	Sum	24,300,000
(5)	General & Administrative Expenses (<u>10</u> % of (4))	2,430,000
(6)	Fee (<u>10</u> % of (4))	2,430,000
(7)	Total Cost	29,160,000

listed. Plane fare, taxi, car rental, and hotel and meal costs should be listed and totaled. Personnel charges are not included if included in the Personnel Cost sections.

9. *Overhead, General and Administrative (G&A) Cost, and Fee Justification.* The overhead, G&A, and fee charged your customer should be an outgrowth of audits performed by your accountant. Work done for the government will require a special government audit by the Defense Contracts Audit Agency (DCAA). The DCAA will often audit at the beginning and end of a contract. Previous audits can be used to justify present rates.

10. *Subcontract.* Any subcontractor should also supply items 3–9 for inclusion in the cost proposal or be submitted by the subcontracts directly to the government.

10.8 PROPOSAL EVALUATION

Once all the submitted proposals are received, they are distributed to a Proposal Evaluation Board consisting of the customer's employees and consultants who are experts in the technical, management, and cost areas. In large contracts, there is a group of experts for each area. Often there is an advisory council of upper managers or, in the case of the government, consisting of Generals, Colonels, Admirals and Captains.

Table 10.7-4   Matrix of Personnel Cost and Hours by Quarter

Personnel	1st Quarter				2nd ... 8th Quarter	Total Cost	
	Hours	Unloaded Hourly Rate	Unloaded Cost	Loaded Cost*+	...	Unloaded	Loaded
W.M. Head	265	$40	10,600	25,440	...		
W.L. Second	160	30	4,800	11,520			
P.Q. Third	300	30	9,000	21,600			
A.J. Fourth	520	40	20,800	49,920			
C.Q. Fifth	320	40	12,800	30,720			
TOTAL	1565	——	58,000	139,200			

* Loaded Cost = Overhead $\times$ (G&A + Profit) $\times$ Unloaded Cost

+ Loaded Cost for consultants usually does *not* include overhead unless the consultant uses your offices. Even then, a reduced overhead is often employed since the consultant does not receive fringe benefits.

Table 10.7-5 Matrix of Personnel Cost and Hours by Task

Personnel	Hours	Task A Unloaded Hourly Rate	Task A Unloaded Cost	Task A Loaded Cost	Tasks B...E		Cost Unloaded	Cost Loaded
W.M. Head	120	$40	4,800	11,520	...	...		
W. L. Second	120	30	3,600	8,640				
C.Q. Third	300	30	9,000	21,600				
A.J. Fourth	200	40	8,000	19,200				
C.Q. Fifth	200	40	8,000	19,200				
TOTAL	940	—	33,400	80,160				

313

Table 10.7-6 Matrix of Task Cost and Hours by Quarter

| Task | 1st Quarter | | | 2nd ... 8th Quarter | Cost | |
	Hours	Unloaded Cost	Loaded Cost		Unloaded	Loaded
A	940	33,400	80,160	...		
B	625	24,600	59,040			
C	0	0	0			
D	0	0	0			
E	0	0	0			
TOTAL	1,565	58,000	139,200			

Usually the Technical Proposal is the most important, with Cost second and Management third. When being evaluated, the Technical Proposal may be divided into areas such as:

1. Soundness of approach.
2. Understanding of the problem.
3. Previous experience.
4. Personnel.
5. Scheduling.

The Cost Proposal is divided into areas such as:

1. Reasonableness of total cost.
2. Matrices.
3. Completeness.

The Management Proposal is divided into the following two areas:

1. Organization.
2. Ability to track and control a project of the type proposed.

These areas differ slightly depending on your customer. If the government is the customer, each area is rated as *superior, acceptable, susceptible of being made acceptable,* or *unacceptable.* An unacceptable rating in any area usually results in disqualification.

If the highest rating of the Technical Proposal is obtained by several bidders, usually the lowest-cost bidder will win.

10.8.1 Best-and-Final Negotiations

Everyone likes a bargain! And the customer is no exception! Before making an award, the procurement officer will call each bidder who remains in the competition and explain what areas the proposal reviewers would like clarified. This may include clarification in all three proposals. Further, the procurement officer may question the fee, the G&A, the salary of an individual, and so on, sometimes to the point of suggesting that the cost might be reduced by some amount.

It is good practice to heed the recommendations of the procurement person. Most likely similar comments were also made to the other bidders.

10.8.2 Debriefing

If you should lose, do not be disheartened. There is only one winner! However, do request a *debriefing*. A debriefing is a meeting with the customer's technical and procurement personnel, at which time they will explain in what manner you could have strengthened your proposal.

Ask about your technical, management, and cost proposals. If they were not rated superior, go over each section with the customer's representatives to determine how you could have improved the proposal.

10.9 CONCLUSION

Proposal writing to *win* begins well before the RFP is released. It is hard work to write a proposal. It is very exciting to win. Then, on to the next proposal!

EXERCISES

1. An RFP is received. Marketing has estimated that the winning bid will be about $5M.

   You are Director of Marketing of a moderate sized company. After meeting with the Director of Engineering you learn that only one person in your company has the experience to write the Technical part of the proposal. You also know that although there is no "favored" bidder, five (5) Fortune-500 companies are bidding.

   List reasons *for* and *against* bidding on this contract. Discuss the possibility of "teaming".

2. Large companies sometimes form *joint ventures* to bid large contracts. Explain why a joint venture is more advantageous than a subcontract

agreement. Why is the joint venture complicated. If you were your company's VP in charge of a joint venture, what would you require in an agreement with the other company.

3. To respond to a given RFP, 5 technical people will be needed for a two month effort.

   a. Who else will probably be required to write the proposal? How many personnel hours would you estimate for each person?

   b. Using your experience, estimate the cost of writing the proposal. Include all indirect costs in your estimate.

   c. What should the minimum contract cost be estimated at?

4. Refer to the Personnel-Experience (P/E) Matrix shown in Table 10.5-2. If you work for a Company, select 10 people that work with you and fill out a P/E Matrix. If you are a student or consultant consider a project of your choice. Create 10 people (including yourself) and prepare a P/E Matrix.

   After completing the matrix, review it as an impartial observer would. Look for strengths and weaknesses. List them. In proposal preparation, these weaknesses should be *rectified*.

5. Refer to the Matrix of Labor and Hours (L/H) shown in Table 10.5-3. Assume that a project lasts for 5 years and that one person-year is 2,000 hours. Task A is to be completed in 9 months, B in 15 months, C in 18 months, D in 1 year and E in 6 months. Also assume that each Task is done in sequence and hours are distributed uniformly for each Task. Complete the L/H Matrix.

6. Explain why Table 10.6-1 is of importance to any customer.

7. Refer to Table 10.7-3. If Direct Costs on a contract are estimated at $1.2M, Consultants at $200K, Equipment at $100K, Travel at $20K, Subcontracts at $250K, and Overhead at 100%, G&A at 10% and Fee at 9%, calculate the Total Cost of the contract.

8. Refer to Table 10.5-3. If Overhead is 150%, G&A is 5% and Profit is 8%, complete Table 10.7-4.

9. Refer to Problem 8 and complete Table 10.7-5.

10. Refer to Problem 8 and complete Table 10.7-6.

11. Refer to Problem 8 and complete Table 10.7-3.

CHAPTER 11

PROJECT MANAGEMENT SOFTWARE SYSTEMS

11.1 INTRODUCTION

As we saw in Chapters 4 and 5, the computational complexity required to analyze progress and produce scheduling and tracking reports makes manual computation very difficult. As a result, engineers and managers recognize the need for and the advantages of project management software.

Effectively, project management software began in the mid-1960s to replace the tedious and repetitive task of manually calculating and recalculating project data. Since then, project management software has evolved over the past three decades from cumbersome mainframe management information systems, accessible only from corporate Data Processing (DP) Departments, to powerful desktop data base systems. Today's software provides individual project specialists much more power for real-time number crunching along with feature-rich options for project cost control and custom management reports. In fact, every Project Manager and engineer now has the ability to access project data, update it, and pass it along to the project office. Even for the very smallest projects, a plethora of applications software packages exist. Written especially for personal computers, these new programs allow development engineers easy data entry and project tracking without additional support from project software specialists or the DP Department.

The demand for increased control of project schedules and costs provides a catalyst for the development of computerized project management information systems. These systems, in fact, will most likely become the most prevalent application programs on the market well into the next decade. Business executives and engineering professionals must therefore determine (1) when it is most advantageous to employ computerized programs to save time and effort over manual techniques and (2) what programs might best suit their requirements to plan, schedule, and control their specific R&D projects.

This chapter explores the key features of a wide range of different project management programs. Emphasis is further placed on developing a systematic methodology for actually choosing a software program while considering current and future data processing needs and the necessity to tie the program to other management information systems.

11.2 WHEN A COMPUTER IS NEEDED

Despite the numerous advantages of using computer software to assist Project Managers and specialists in planning, scheduling, and controlling their R&D projects, a computer is not required or desired for all applications. Computer processing involves time and extra cost to implement. First, one must have ready access to a computer before purchasing the program and finally entering project data into it for analysis. For small projects, the time required for data entry may preclude the use of software. Furthermore, a minicomputer may be required to run the applications programs best suited for very large projects. The complexity of these sophisticated software programs necessitates the use of specially trained personnel, which adds considerably to the cost. In general, the more useful the software, the higher the cost and the less user friendly thereby requiring the need to dedicate project staff to operate it. Thus, the choice of software one employs to generate and update project networks or cost and schedule reports must be made carefully.

Hundreds of software packages are available for almost every make of minicomputer and personal computer. Each system offers numerous advantages that by far outweigh the tedious task, for instance, of manually calculating and recalculating an early- or late-finish activity schedule.

There are several reasons why computerized management control benefits a project. These are outlined below to help to determine when it is most advantageous to employ a computer to process project data.

1. *Time-phase the project baseline plans.* Computer software can ease the work required to formulate multiple iterations of project baseline plans as well as document the initial planning process for further reference.

2. *Generate the Project Work and Organizational Breakdown Structures (WBS/OBS).* Chapter 10 emphasized how on-line storage of project plans is often required by the Project Contractor. Quick and easy access to these files, whether for consultation, review, or updating, is a must. Graphic portrayal of both project structures also proves to be a helpful tool. For example, the DOD and other government contractors

usually stipulate that both the project WBS and OBS appear as exhibits in a Request-For-Proposal (RFP) response. Projects that share resources from common pools also benefit from graphic portrayal of the two structures because the common work packages are easily highlighted.

3. *Compute the Project's Schedule and Early and Late Finish or End Dates.* R & D projects comprised of as few as ten activities will benefit from a software solution. But on efforts where the number of project activities is several hundred, the storage and retrieval by computer of project data leading to these calculations is a necessity! In addition, in the typical R & D, multiple project environment, there is generally a need to create:
   · Subnetwork schedules of project activities that roll up into a subsuming hierarchy for processing separately or as an integral subset of the master network,
   · Networks that require multiple start-up dates,
   · Schedules that are computed based on assigned times of different durations (i.e. hours, days, weeks, months, etc.) worked by individual project resources.

4. *Centralize Project Data Collection.* For project specialists who reside in geographically dispersed plant locations, centralized maintenance of project schedules and cost reports is essential. This is also true for project personnel who work simultaneously on a number of different small- or medium-size projects. By centralizing project files, project personnel can electronically transfer, between corporate data bases, information needed from other DP or MIS systems (i.e., corporate accounting).

5. *Track and Update the Project Schedule.* Both the master schedule and cost variance reports require frequent and periodic updating to help to determine when project work is to be completed and at what cost. Changes, because of their frequency of occurrence, can't be readily incorporated into these reports unless a computerized processing system is available.

6. *Search, Sort, and Display Project Data.* Searching for and sorting project data constitute one of the primary tasks the Project Manager is concerned with in order to generate timely and meaningful project reports. Because project control is accomplished by analyzing a variety of plans and reports, the output capacity of computerized project management systems helps to direct this management activity. Production of graphic reports of the project plan is another valuable tool provided by computerized programs.

7. *Forecast and Manage Project Resources.* Most medium- to large-scale R&D projects have finite but defined resources (personnel, equipment, materials, etc.) that must be scheduled and tracked against

original plans and budgets. The capacity to code resources assigned to specific activities and tasks facilitates their multiple allocations to a variety of different projects. Another important requirement is the ability to smooth out or level resources used over time. Unpredicted constraints may therefore be accounted for by simple automatic adjustment of the project schedule.

8. *Integrate, at the project manager's fingertips, a number of sophisticated techniques for advancing the flow of project information across engineering and manufacturing sites.* Combining on office workstations project planning and scheduling systems with Computer Aided Engineering (CAE) and Materials Requirements Planning (MRP) software would satisfy Engineering's simultaneous need to monitor design changes and to track procurement of parts and materials on the factory floor while scheduling shipments of completed systems.

11.3 KEY FACTORS TO CONSIDER

Most project management software packages offer managers and project team members flexible control of project plans, schedules, costs, and resources. In fact, the recent proliferation of these programs requires that the project's various management requirements be carefully considered before final selection and implementation of the chosen program. Most importantly, it is the user's needs that dictate what software is selected, not the reverse.

11.3.1 Considerations for the Software

As a preliminary step in actually selecting a computerized management system, one should analyze management's exact expectations and detail them in writing. This way the features, capacity, cost, and limitations of the considered software programs will be readily available for review. In your analysis consider answers to the following list of questions. These questions will help you concentrate on the key factors to consider when selecting a computerized management system:

- What features are offered that will enable your team to plan and to control project activity? How does the program handle the creation of project plans, resource allocation, the optimal use of scheduled activities, and the comparison of actual versus planned progress?
- What constraints, if any, does the system impose on the size of the project network?
- What standard presentation graphics are included in the program? Do they illustrate the full integration of schedule, resource, and cost

information as well as denote the dependency relationships among project activities?

- What level of quality is offered for presentation graphics? Can network (precedence) diagrams be generated directly without the need for additional programming, specialized sorting, or coding of project information? Are there provisions for displaying the network plots by activity names and WBS/OBS codes or for specific time intervals? Can the graphics be interpreted easily by senior management without the aid of a professional planner or DP consultant?

- Does the system generate easily accessible, on-line printed or plotted planning charts? Can professional planners and project team members obtain a plan baseline without having to produce presentation–quality charts?

- Are project team members allowed the fexibility of working interactively with the system to manipulate project data while ensuring the integrity of the project's plan?

- How difficult is the software to learn and to use for both novice and experienced project personnel? Is specialized staff needed as a support service or to train other engineers and project personnel? Is the program's documentation straightforward? Is its menu structure easy to operate? Can user-specific, data-entry screens be tailored to match the different requirements of a variety of projects? Are command-driven sequences and program macros available for fexible data entry as program users become more sophisticated?

- What are the initial installation charges? How much is the required capital outlay? Can available computing hardware be employed, or is specialized equipment required? Are there hidden charges for continuous vendor troubleshooting, hot-line support, or vendor-conducted training? Is this support available throughout the life of the project?

- How frequently does the vendor update the product? Are product enhancements for new system releases or documentation updates supplied free of charge?

- Will the software operate on more than one vendor's hardware?

- What printers and plotters can be used with the proposed system? Are sophisticated laser printers supported? Can these peripheral devices be located at sites remote from the computer?

- Can the user interface/link additional MIS programs, publications software, spreadsheets, and databases with the system? How easy is it to import/export project data to and from the project management system and these programs for recalculation and presentation?

- Does the system under investigation support an open architecture for horizontal integration of data from corporate CAE systems, on-line inventory control or order-entry systems?

- Is the software flexible enough to accommodate demands for additional processing capacity, increased sophistication in the types of reports requested or the number of projects tracked? Can today's system grow to handle tomorrow's activities?

11.3.2 Considerations for the Hardware

The next step in this analysis is to consider the availability and distribution of the computing hardware to be used. Thorough assessment of presently employed computing resources will help the Project Manager to weigh the implications of using various computing configurations for implementation of the selected project management software.

Certainly there are considerable advantages in using embedded hardware to manage projects in a time-sharing mode. These include allowing project team members remote access to a centrally located host. Today, this can also be done via local area networks (LANS) that link together several personal computers. This type of arrangement may be extremely important during the early life-cycle phases of large R&D projects, for instance, when production of feasibility studies must be generated before funding is committed.

Considerable flexibility is also gained by choosing software that operates on computers of more than one size. Both large and small projects can be managed with project software designed to run on families of host computers that use the same operating system. This facilitates the tracking of the simplest to the most complex project efforts. Project team members need learn the specifics of only one selected software program that apply to their own designated project management tasks.

Today, new communication software is being released for commercial applications that enables the exchange of files between different operating systems over communication networks. The networks are able to link the smaller personal computers with one another as well as the more powerful minis. Such an exchange of data, for example, allows managers and project teams to share critical information regarding project resources that was entered first into an MS-DOS* project management program and must be integrated into other corporate files that are UNIX-based® data systems.

The choice of computing hardware should fit the project structure. Likewise, the tasks associated with system administration need to be accounted for and assigned to either project personnel or a corporate DP support organization.

Cooperation of the software vendor is key in determining:

---

* MS-DOS is a registered trademark of Microsoft Inc.
® UNIX is a registered trademark of AT&T Bell Laboratories.

- Who will have access to project files.
- How the software system is to be installed along with back-up file control and security.
- Whether to link and/or configure the software for data exchange between any existing management systems in operation.

It is also important to determine whether the potential vendor will service the project management system once it is installed, or make available skilled programmers and analysts to extend, tailor, and further develop application programs at its customer's request.

It is also important to consider whether the vendor supplies remote diagnostic assistance or on-premises maintenance and repair of either the software system or any turnkey devices.

11.4 FINDING A PRACTICAL COMPUTER SOLUTION

Fundamental knowledge of the concepts of project planning, scheduling, and control is an important prerequisite for satisfactory program selection. The advantages offered by one system over another are sometimes subtle. Therefore, the manager or engineer embarking on a comparative shopping spree can best judge the benefits provided by a prospective system when he or she understands the tasks the software must perform and has a determined perference for how they ought to be performed. Also, it is important to have realistic expectations because not all the software available commercially is sufficiently sophisticated to meet every manager's requirements or even support all the analytical techniques espoused in this textbook. Despite prolific feature improvements, experienced R&D managers report very real system limitations. So knowing your own project management requirements is key to making the best choice. As discussed in the previous two sections of this chapter, there are a number of variables to weigh and to thoroughly consider no matter how small or large your project effort and no matter the magnitude of the software system under investigation.

11.4.1 Comparing Systems and Their Features

To acquaint prospective users with the variety of software solutions, the next section in this Chapter compares the features and capabilities of many of today's commercially-marketed programs. Specific limitation imposed upon system users and other common deficiencies are also described.

Very large and complex projects with R&D life-cycle budgets upward of $10,000,000 justify the investment of $100,000 to $200,000 for a powerful mainframe program to coordinate data collection as well as centralize the planning, processing, and control of project information from project initiation through postmortem evaluation. Such programs also require the use of specially trained personnel that further increases the cost of using project management software.

Although the forecasted demand for expenditures for micro-based project management software packages is expected to grow from $28 million in 1985 to $82 million by 1990 [1], today there is a steady $100 million market for mainframe- and minicomputer-based systems. The majority of programs available for these computers are comprehensive and offer high-speed processing power. Most minicomputer software systems are extremely flexible, portable, and range in price from $20,000 to $150,000.

The 50 or so microcomputer-based packages on the market offer innovative data entry and, for the most part, extreme ease of use. (Refer to Appendix 11.1 for a listing of the more popular personal computer (PC) programs and their vendors.) Some of these systems were ported from their more powerful mainframe forerunners and are scaled-down variations of the original programs. Typically, these micro packages require higher-end computer configurations such as a 10-megabyte hard disk and extended memory of more than 360K. They generally are priced from $2000 to $5000.

The majority of PC project management programs selling between $200 and $2000 are purchased most often by corporate Data Processing and Management Information System (MIS) Departments needing quick and easy access to project data and information for several small- to medium-size project efforts, all fairly short in duration. [2]

Assessment of the value of these programs must be based on a comparison of how well the software's features enable project managers and engineers to plan and control their project efforts and costs. A program's capacity and performance can be judged adequately only in relationship to how well the software system enables the management team to:

- Measure over time the cost of realizing project objectives,
- Plan and assess progress,
- Allocate resources,
- Produce key reports.

The degree of friendliness, sophistication, and power the program possesses is an important requirement dictated by the tasks at hand and the magnitude of the project effort.

11.4.2 Data Entry and Manipulation

The more popular PC project management packages in the $200–$2000 price range support electronic, interactive data processing and production of project plans and schedules right on the computer screen. For instance, *Harvard Total Project Manager* builds its "road map" of project activities on the screen. In addition, managers working with *MacProject* would use the Mac mouse for simultaneous creation and display of project flow charts that resemble precedence diagrams.

Other programs, such as *Microsoft Project, PMW,** and *Time-Line,* construct electronic Gantt charts of the project schedule, while *Super-Project Expert* displays the schedule of project activities and tasks as a precedence diagram that can be manipulated as it builds in monochrome or color. Data entry is accomplished in the popular program, *ViewPoint,* by building the project Work Breakdown Structure graphically on the screen.

A number of the micro-based products priced upward of $2000 (e.g., *Artemis Project, Primavera Project Planner, PROMIS, and Vue*) employ menu-driven data entry screens common to the more powerful mini- and mainframe project management systems. Data entry is interactive. Historical data is retrievable for easy updating, validation, error checking, summarization, and/or reporting. Project bar charts, networks, and logic diagrams may be previewed on the computer monitor before plotting or printing.

Another feature of these types of programs is the capacity to combine and consolidate data from multiple projects and upload it to bigger computers for more rapid and efficient processing. Subproject data, entered locally by a project engineer, can be sorted, previewed, and modified before it is centrally processed along with other project information at the Project Office, where multiple project progress will be analyzed and subsequently reported on.

Minicomputer-based project management software and mainframe systems (see Appendix 11.2 for a listing of the more popular vendors' programs) operate on a variety of different-sized and different-priced computers sharing identical operating systems. Data is entered, typically, through client-customized data input screens and, depending on the software system's individual features, may be stored for unlimited retrieval in either a hierarchical or relational Data Base Management System (DBMS).

The *APECS/8000, Artemis,* and *Vision* project management programs, for instance, feature a modularly integrated DBMS that functions independently of the computer operating system. This provides project personnel direct access to all project data through the DBMS and also permits unlimited retrieval for manipulation and reporting. New fields of data can be appended or deleted at will, allowing for lengthy "what-if" analyses of project schedules, costs projection, and resource deployment. Baseline plans stored in the DBMS would not be altered during such analyses.

---

* Project Manager Workbench is a trademark of Applied Business Technology Corporation.

These programs also provide the capability for making global and conditional changes to the database to generate, for example, schedule "updates" or other requested progress reports.

11.4.3 Scheduling Techniques and Graphing Capabilities

One of the most important considerations when comparing project management software systems is to determine whether the software's capacity for establishing a project schedule meets your own project requirements. Although they offer enormous variations in both on- and off-screen project road maps, dependency diagrams, traditional Gantt charts, and milestone schedules, very few of the popular, commercial programs generate networks calculated using P-PERT, or the three-time statistical schedule, discussed in Chapter 4. At this writing, the authors have found only one tool, *Super-Project Expert,* that performs P-PERT. SCS Telecom's tool, *Prob-PERT,* operates in conjunction with the computer spreadsheet program, *Lotus 1-2-3,* to perform the probabilistic calculations. And while at least one PC-based program, *PERT Master,* for instance, generates arrow as well as Activity-on-Node (AON) precedence diagrams, most do not. *Harvard Total Project Manager* and *Microsoft Project,* for example, use only the AON technique.

Other PC packages restrict network size by limiting the total number of activities that might be assigned for processing at any one time. Some of these programs will handle only limited numbers of networks per copy. Many do not allow full integration of schedule, resource, and cost information across projects that share common resources. While most make available multiple work-week calendars, the maximum calendar length for a project typically cannot exceed several years.

Some of these packages also constrain the user's ability to impose multiple start and end dates on project activities and, therefore, to compare planned versus targeted early- and late-start schedules. Subnetworking and the ability to roll up resources and costs from multiple subnetwork levels may not be allowed. Some programs may fail to verify network logic before attempting to produce a schedule. Such validation is essential for accurate project planning and control.

It is also important to assess the software programs' capability to produce presentation-quality network diagrams and bar charts. Graphic displays, whether printed or plotted directly from the PC screen or the scheduler data base, need to be clearly identifiable and easy to read by both project personnel and upper management. Truncated, or poorly abbreviated, graphic headings, along with unsorted data displays can be easily misinterpreted. This could cost conscientious managers or engineers their project assignments.

Virtually every medium- to large-scale mini- and mainframe computer vendor offers a standard graphics scheduler. Most provide, in addition, optional x–y-plotter modules that display, in color, management graphics and cost–schedule control plots.

These programs typically enable users to define their own plot zones on the basis of preselected activity codes or other alphanumeric descriptors. Users are allowed, for instance, to narrow the focus of a network display by (1) specifying precise time slices, (2) sorting ranges of targeted or flagged activities, (3) selecting data that correspond to particular work breakdown or organizational breakdown codes.

Bar chart displays generally include both rectangular and traditional line or triangle notations. Generally, sort sequences and user-defined summations can be retained in a library for subsequent regeneration. Most programs also provide for the integration into text of user-defined symbols. These can be placed near the actual bars or be displayed in a legend.

11.4.4 Resource Management

Project management PC programs selling within the $200–$500 range generally allocate resources to project efforts by assigning them to specific project tasks or activities. *Microsoft Project* Version 4.0, for instance, allows up to sixteen different resources to be assigned per project task right on the Gantt display screen. Resources identification requires specifying, for even fractional units, the number of resources needed to perform each activity. To maintain control, resources must be monitored across multiple projects. Because resources are linked automatically to project schedules, this program enables managers to track their allocation over time and print out for individual team members handy listings of work assignments and associated start and end dates.

Another popular PC program in this price category is *Time-Line*. *Time-Line* flags in a split-screen display overextended resources that may be leveled to redistribute the overload across project activities. The popular *SuperProject Expert* displays resource conflicts in its Resource Details screen. Here the Project Manager is permitted to resolve conflicts over personnel scheduling by leveling resources by preassigned priorities.

Mid-price-range programs such as *Project Manager Workbench, Pertmaster,* and *ViewPoint* maintain a full data base of project resources. Resources can be allocated and tracked by resource type (personnel, capital, equipment, etc.). Conflicts regarding availability or usage within a project or across multiple projects are resolved through automatic scheduling and leveling. If a project must be completed by a specific "drop-dead" end date, the manager may selectively overload resources to meet that date.

The high-end and more powerful PC and minicomputer systems assign specifically available resources to project activities from general classification pools (e.g., Software Engineers, Hardware Engineers). During the project's life cycle, resource availability varies on a daily and even hourly basis. When the number of resources required to perform particular project work is critical to its successful completion, their utilization must be tracked consistently across assignments by (1) resource type and (2) project activity code.

Systems such as *Mister* and *Vision* produce resource constrained schedules and reports that display this information for the personnel assigned to priority activities for specific or multiple projects. Resources can therefore be leveled against user-assigned priorities for optimal deployment.

11.4.5 Cost Management and Performance Measurement

Any project endeavor, whether commercially based or government-sponsored, requires application of good cost management techniques. While the most sophisticated mini- and mainframe systems allow managers to produce mandated DOD, DOE, and NASA cost–performance reports, many of the smaller PC-based packages excel in comprehensive budget and cost reporting.

Microsoft Project is a basic, entry-level system that allows cost aggregation and tracking of resources assigned to project activities. Cost data is entered at the same time resources are specified on the program's Gantt display screen.

Another project manager system, available for both MS-DOS and UNIX/ XENIX computers, is *Microtrak*. *Microtrak* dynamically schedules project resources while automatically tracking their costs. Cash flow histograms and resource cost reports are generated as standard reports.

As projects become increasingly complex, the ability to store cost data and generate custom reports from this data base is subsequently advantageous. *SuperProject Expert* is most adept at managing project costs. Total project costs are calculated from processing data entered on the Project Details screen. Aggregate costs are further broken down for each task and project resource. This data is accumulated on input at the program's Task Details and Resource Details screens. *SuperProject Expert* allows for manipulation of unique fixed and variable costs as well as actual, planned, or budgeted costs assigned to either the entire project, complete tasks, specific resources, or individual team members. This package allows managers to apply costs when tasks begin or end and to prorate them over a project's life cycle.

The *Advanced Project Manager Workbench* also provides cost management. This program tracks variable as well as fixed costs at the project phase, activity, or task level, and for individual resources. Projects are viewed from multiple screens that display simultaneously the Gantt chart and resource spreadsheet. Included among the tracking reports are a work effort and cost status report.

PROMIS and *Primavera Project Planner* are high-end microsystems that enable managers to track the costs of fluctuating resource usage over the life of a project. Cost accounting is automatic in that the costs of resources required to accomplish project work are updated whenever new schedules are generated.

Primavera produces three different time-phased schedules, including a baseline and schedule–resource–cost plan. These may be used to measure the project's earned value. Other features include the ability to project activity costs and roll up or summarize expenditures in accounts from any level of the Work Breakdown Structure (WBS). Graphic and tabular cost reports may be displayed either on the computer screen or as printed or plotted hard copy.

PROMIS defines costs through project resources and through project activities. Cost estimates and actual cash flow may be summarized for each activity or resource by different daily, weekly, or monthly periods as well as for targeted and actual schedules. *PROMIS* also provides easy consolidation of cost and cash flow data from multiple projects. The system coordinates subnetworks, minimizing problems that would arise from conflicts among multiple cost or resource codes when projects are merged into one network. *PROMIS* also gives managers the capability to generate multiple schedules (early and target) and multiple cash flow reports (periodic and cumulative) for individual resource profiles (e.g., all hardware engineers, all software engineers).

One advantage of employing sophisticated mini- or mainframe software systems to manage project finances is their enormous capacity for monitoring and measuring actual costs versus established budgets over the complex and ongoing life-cycle phases of very large project efforts. Many of these programs can be configured to accept corporate financial data to establish activity costs incurred on the project. They may further allow summarization by accounting codes that can be sorted in any sequence determined by project personnel. Some also make available integrated cost/schedule control and multiple project analysis for projects that share common financial resources.

Ideally, budgeted estimates and project cash flow are updated automatically each time the schedule is altered. In general, cost and schedule variance as well as variances to the original project baselines are obtainable for any level of the project.

Some of these systems also produce the standard DOD, DOE, and NASA schedule performance series 7000.1 and 7000.2 reports. One word of caution—many of these products' claims exceed the real capabilities of the systems to reach C/SCSC (Cost/Schedule Control System Criteria) compliance.

11.4.6 Communication

Inevitably, the Project Manager who works with one popular microcomputer-based software program will want to transfer or upload or download project data to another system for additional processing. The popularity of the new 32-bit microcomputers have made it advantageous for many vendors to provide their software users with this built-in capability.

Many PC-based packages are capable of communicating with larger systems. *Vision-Micro* and *Artemis Project* accept, share, and transfer data between both mini- and mainframe computers running, respectively, *Vision* or *Artemis* software. Likewise, *SuperProject* uploads to *CA-TELLAPLAN*. As the project team's expertise and project management requirements grow and change over time, so, too, can the computing power needed to plan, schedule, and control project work.

Other vendors have begun to collaborate on linking their systems in a manner that will allow data to be exported without reentry into more advanced systems. Data exported from *Microsoft Project* is transferable to *Primavision* through a special feature conversion program. This eliminates the need for *Microsoft Project* users to employ other PC graphic programs to create plotted, presentation-quality management graphics or charts.

Project management programs such as *Project Manager Workbench, SuperProject Expert,* and *ViewPoint* allow file transfer to such popular systems as *Lotus 1-2-3, Super Calc, dBase,* and many other information processors, toolkits, and graphics packages. *Workbench* even reads files created by other programs in standard PRN format.

11.4.7 Decision Support and Expert Systems

Managers of very large project efforts undoubtedly will find computer-aided decision support beneficial. In addition to multidimensional capacity for data manipulation, retrieval, and display, these problem analysis, expert systems are a computing resource for trading off design alternatives, scheduling options, resource availability, and project cost estimates with actual performance measures.

These programs access data that may be stored in large corporate data bases and provide direct interface to the project's operational management system. Expert systems build project scenarios by modeling specific problems to be solved. Project personnel are presented with a series of questions, graphic displays, and summaries that demand interpretation to give conceptual meaning to the data. Thus, a human-computer dialog is carried forth where the manager asks for and responds to simple computations, comparisons, or projections as a means for providing input back to the support system. The computer then models its solution to the specific problem much like a human expert does when reasoning among alternative variables.

There are multiple advantages for using commercial decision support or expert systems to perform decision analysis:

- Most make available probabilistic techniques for determining possible overages and slippages.
- All build a knowledge base of the project that defines the problem

according to a set of predefined rules in order to draw solution possibilities that can be called up at random by project personnel.

- Data manipulation is performed through the use of natural query languages.

To conclude this section, refer to Table 11.4.1 for more on the many features and attributes to look for in project management software systems. Frequently experienced problems by R & D users are enumerated in Table 11.4.2.

11.5 INSTALLING A SYSTEM THAT'S RIGHT FOR YOUR PROJECT

Given the variety of computing hardware and software systems available for efficient management of project information, the approach taken to select the most effective hardware–software combination must be systematic and methodical. Before embarking on your management's charge for selecting one or more of these systems to automate project planning, tracking, and decision-making, be sure that you have thoroughly analyzed *all* appropriate management requirements as well as determined what size computer is needed to run each system under consideration.

Begin by reviewing your written list of criteria to make this decision, in other words, your management's and the potential users' particular planning, scheduling, or tracking applications the software must perform. If management is satisfied with your analyses, then proceed with a discussion of how to configure the various systems within the project office or the corporate DP Center. This lets you decide whether your choices impose any difficulties. Then determine, as accurately as possible, what each program will cost to operate, and compare these charges to the projected performance savings.

All this should be done while reading current articles about the software systems found in trade journals.* Then, to help ensure that the software you're considering has the potential to meet all your needs, contact specialists involved in project management activity and conduct surveys among other corporate planning and project offices. Spell out your requirements in writing, so that the features, capacity, costs, and limitations of several potential systems that you have selected can be studied in more detail. Ultimately, you'll narrow down the list of accessible systems to a

---

* *InfoWorld, PC Magazine* and *PC World* periodically publish ratings of MS-DOS based software systems. The *Project Management Journal,* a bi-monthly publication of the Project Management Institute, features regular reviews of different-size systems. Additional references concerning the applications and limitations of these systems are recommended at the end of the Chapter.

Table 11.4-1 Important Attributes/Features to Look for in Project Management Software Systems.

Data Entry	User Interface	Data Manipulation/Communication	Scheduling Techniques
1. Multiple menu screens or pop-up menu windows for easy data input	1. Optional mouse-driven interface	1. Automatic retrieval and display of historical project data when updating existing data	1. Critical Path (CPM) scheduling with availability of the following:
2. Command-driven sequences that allow users to operate the system with the menus turned off	2. "Cut and Paste" commands that let users compose or draw project displays right on the computer screen	2. Ability to preview reports and plots on the computer screen prior to printing or plotting	· AOA and AON, linear and variable, time-scaled networks
3. Full-screen forms that users can tailor for customized applications	3. On-line help facilities, interactive tutorials, and user training that expedites learning	3. Ability to save reports to disk and read and write files in standard communication formats	· Probabilistic PERT networks with milestone notations
4. Dedicated function keys that replace libraries of commands for all needed functions	4. Complete error checking and data validation	4. Project import/export to other project management systems, publication software, spreedsheets, etc.	· As-soon-as-possible and as-late-as-possible activity-based schedules
5. Automation of repetitive procedures through programmable subroutines or macro commands	5. Copy commands that let users save project data for later reuse or move portions of the project elsewhere	5. Access to an independent Data Base Management System (DBMS) for retrieval, manipulation, and reporting of project data	· Early start/early finish and late start/late finish activity reports
	6. Features that provide for automatic display of subsequent levels		2. Multiple work-week calendars and maximum calendar lengths for projects that spand several years

3. Flexible coding structures that allow users to code project tasks by WBS/OBS numbers or alphanumeric descriptors

4. Progress reporting and full summarization of scheduled activities by:

   · User-defined status dates

   · Actual start

   · Actual finish

   · Duration complete

   · Duration remaining

   · Percent complete

5. Multiple-level project processing and provisions for linking project activities for multiple project reporting

6. Ability to interface with other corporate DP/MIS systems

of detail in the project plan, WBS, or activity sequences

Table 11.4-1 (*Continued*)

Resource Management	Cost and Performance Measurement	Graphics and Displays	Other Features
1. Ability to schedule multiple projects from a common resource pool and to monitor changes in resource availability on project completion dates and costs	1. Automatic conversion of resource usage to cost	1. Availability of the following: · Work breakdown structures	1. User-definable options and reports
	2. Full integration of resources and cost with schedule	· Network diagrams	2. General and context-specific help facility; easily assessible and informative error messages
2. Ability to level resources by resource availability and override constraints imposed by time-limited activities	3. Cash flow analysis 4. Detailed budgets by activity and resources	· XY charts · Bar (Gantt) charts · Logic diagrams · Resource histograms	3. Vendor support and maintenance agreements that include:
3. Capability to level resources for a single resource class, particular activity, or the entire project	5. Reporting by: · Actual costs to date (ACWP) · Estimate at completion (EAC) · Budget to date (BCWS)	2. Ability to display multiple curves, histograms, and plot points on the same graph to determine variances and predict trends 3. Provision to edit and generate multiple-size plots	· Telephone hotline consultation · User training · Documentation updates

4. Ability to display resource usage histograms directly on the computer screen to assess the effect of variable resource availability over time

· Budget at completion (BAC)

· Earned value (BCWP)

6. Summarization of cost data across multiple projects

4. Ability to specify sort sequences by activity name or WBS/OBS codes

5. Ability to retain in a library or file user specifications for subsequent regeneration of bar charts, network diagrams, etc.

6. Device-independent interface for printing and plotting graphics and other displays

7. Ability to summarize and selectively print sections of the bar charts and network diagrams for top-level reporting

8. Ability to integrate text and user-defined symbols into the graphics

· Maintenance and repair

4. Support services for very large and expensive systems that include:

  · Applications program development

  · Consulting contracts

5. Companion products:

  · Optional plotter graphics system

  · Mouse support

  · Communication software for networking several systems

Table 11.4.2 Commonly Experienced Problems with Commercial Systems

DATA ENTRY	· On-screen production and manipulation of the project network on small systems can be quite cumbersome.
	· Inability to assign user-specified activity codes in any size system makes tracking similar activities for different WBS levels difficult.
	· Absence of input procedures that provide users with an alternative to menu-driven screens limits program application to smaller projects.
	· Software configurations for personal computers sometimes prohibit data input by geographically dispersed users.
SCHEDULING TECHNIQUES	· Limitations as to the types of precedence relationships or lead/lag times accepted by the software may result in an inaccurately scheduled project.
	· Confusion over the difference between P-PERT and CPM scheduling algorithms, as well as which methods the commercial systems use, poses difficulties for less advanced users first learning the terminology.
	· Inability to integrate within one software program schedule, resource, and cost information means time and more expense to import/export this data to and from other programs.
GRAPHICS	· Printed graphics that are merely a snap shot of the computer screen require extensive manipulation to generate legible, timescaled networks which are suitable for presentation to upper-level management.
	· Limitation of the number of standard graphic reports or the inability to customize reports is a very real problem. Users should be given enough flexibility for generating as many different kinds of displays as wanted and for designing report formats that meet their *own* customers' needs.

Table 11.4.2 *(Continued)*

SYSTEM DESIGN	· Lack of a fully integrated, modular design creates havoc for project personnel when everything cannot be updated at once. This nightmare is compounded when separate programs are utilized for the storage and retrieval, for instance, of corporate financial records.
DOCUMENTATION	· Poorly written user manuals or performance (job) aids only confuse system users.
	· Overly complex or extremely simplified manuals fail to serve the requirements of either novice or advanced users. Clear and concise, multilevel documentation is sorely needed.
VENDOR SUPPORT	· Purchasers of the less expensive systems are frequently disappointed with the limited support provided by software vendors.
SECURITY	· Password protection and restricted data access, in earlier days, were the domain of the corporate DP department. This issue raises very serious concerns for personal computer users.

possible two or three, whose features and performance you will appropriately assign a weight and then rate accordingly.

What you learn from your colleagues who use these systems will enable you to understand any major drawbacks and to avoid common pitfalls. You might ask to review sample written reports that represent the entire gamut or scope of their project work. Much is revealed by raising such interesting questions as whether the software was ever configured to handle input from other corporate MIS systems. Also ascertain how acceptable the system's standard output is for executive-level briefings. If you intend on making your project management system available on a variety of personal computers, determine how your colleagues handle the problem of multiple access to data files. How secure did they make the computing environment? On larger computers, what protection schemes did the vendor offer to restrict user access to different levels of data? Was it a concern to maintain multiple copies of the system to separate specific, government-funded projects?

You may also want to determine how your colleague's software handles the allocation of common resources who work on multiple projects. How adequately did their systems track resource usage and expenses across these projects? Were they able to generate daily work assignments as well as track their completion? Also, how difficult was it to coordinate user training or disseminate the system documentation? Who provided the training—a corporate computing group or the software vendor? How instructionally sound was the training; is there an optimum number of students that should be scheduled at any one time?

Knowing the answers to these and many other questions will enlighten you about the most commonly encountered pitfalls as well as help you be more realistic about your expectations.

Remember, final selection of a good project management system depends on not only whether but also how it meets your required criteria. Initially, many software systems may appear to satisfy many, if not all, of your general management requirements. For instance, until a more thorough evaluation is made by extensive trialing of one or two demonstration systems obtained from the program vendors, it is not possible to judge the speed at which data is processed or the adequacy of on-line reporting. Benchmark project data taken from similar efforts can be used to evaluate a number of program parameters: capacity, power, versatility, user-friendliness, and graphic output.

Now, use your list of requirements to complete your decision matrix and assess the program features you want to compare. This is accomplished by assigning priority weights to all essential features and then ranking their performance. (A score from one to five is suggested, where a value of five means that the software's performance is outstanding, and a one equals poor performance.) Then multiply the preassigned numeric weights for each feature by your ranking value, and calculate a grand total. As you populate the matrix, it will become evident that, in order for any of the software systems to be selected as candidates for further analysis or purchase, they must receive high scores in your most important categories. Rework the matrix more than once to validate your analysis. This is especially important whether you decide to lease the software prior to full-fledged purchase or install it immediately without further trialing.

Table 11.5-1 details the steps included in this systematic selection procedure and lists some cautions one should take. As you embark on evaluating and installing one or more of these potentially excellent project management software systems, good luck!

11.6 CONCLUSION

This Chapter has explored specific techniques recommended by the authors for systematically evaluating and selecting the right project management software to match project requirements. Also described were a variety of

Table 11.5-1   A Systematic Process for Selecting Project Management Software

Step 1—determine own project management requirements

Who will use the software program?

What limitations or restriction are imposed by the computer hardware? system requirements?

What features are absolutely essential? nice to have in the future but not necessary now?

How does the program's performance compare with its cost to operate?

Who must review the program's output? Are presentation graphics required?

What applications does the software need to perform?

Step 2—search product/buyers' guides to project management software; contact other users to discuss general impressions, limitations, precautions

Step 3—review the many product entries to eliminate programs that fall short of your list of requirements

Step 4—narrow the list of accessible software systems down to a minimum of 2 or 3

Step 5—contact different system vendors to obtain product literature, evaluation kits, or complementary passwords; attend or conduct a system demonstration using a case study of data taken from your *own* projects

Step 6—test each of the features and performance categories for each vendor's system; analyze problems encountered; validate the performance by testing each system more than once

Step 7—compare the results of these evaluations; a performance weighted ranking technique is recommended

---

system features and attributes available today for different-sized and different-priced computers. Cautions that one should take prior to implementing a choosen system were also discussed. More information on the applications and limitations of these software systems is given in the references cited below.

REFERENCES

1. Bannister, H. "Project Software Demand Growing", *InforWorld,* Vol. 8, No. 4, January 1986, p. 1.
2. ibid. p. 1

BIBLIOGRAPHY

Avots, I. "How Useful Are The Mass Market Project Management Systems?" *Project Management Journal,* Vol. 8, No. 3, August 1987, pp. 58–60.

Blum, M. "Incorporating Science and Art Helps Micro Managers Make Software Selections," *InfoWorld,* Vol. 8, No. 39, September 1986, p. 29.

Cobb, J. E., and Diekmann, J. E. "A Claims Analysis Expert System," *Project Management Journal,* Vol. 17, No. 2, June 1986, pp. 39–48.

Fersko-Weiss, H. "Master Plans: Project Management Software," *P C Magazine,* Vol. 6, No. 16, September 1987, pp. 153–209.

Heck, M. "Project Management Programs for Executives," *InfoWorld,* Vol. 10, No. 21, May 24, 1988, pp. 47–59.

King, N. E. "A Decision Support System For Mine Evaluations—A New Tool For Project Planning," *Project Management Journal,* Vol. 17, No. 1, March 1986, pp. 65–81.

"Management and Administration," *ICP Software Directory, Mainframe & Minicomputer Series,* Vol. 3, 59th Edition Spring 1988 pp. 33–47.

Miller, M. J. "Project Managers Still Coming of Age." *InfoWorld,* Vol. 7, No. 52–Vol. 8, No. 1, January 1986, p. 26.

"Office Automation & Business Management," *ICP Software Directory, Microcomputer Series,* Vol. 3, 59th Edition, Spring 1988, pp. 44–59.

Reimann, B. C. "Decision Support Systems: Strategic Management Tools for the Eighties," *Business Horizons,* September–October 1985, pp. 71–77.

Stepman, K. M. *1986 Buyer's Guide To Project Management Software,* Milwaukee, WI: New Issures, 1986.

Tarlson, N., and Yalonis, C. "Convenience or Fuctionality? Hard Choices for PM Users," *Software News,* Vol. 6, No. 8, August 1986, pp. 39–49.

Waterman, D. G. *A Guide to Expert Systems.* Reading, MA: Addison-Wesley, 1986.

APPENDIX 11.1 A SAMPLE OF POPULAR MICROCOMPUTER PROGRAMS

Program	Vendor	Hardware/Operating System
ABT Project Manager Workbench	Applied Business Technology Corp., 365 Broadway, NY, NY 10013 (212) 219-8945	PC/XT, MS-DOS

Program	Vendor	Hardware/Operating System
ARTEMIS PROJECT	Metier Management Systems, Inc., 2900 N. Loop West, Suite 1300, Houston, TX 77092 (800) 362-9118	PC/XT/AT, MS-DOS
ARTEMIS 2000	Metier Management Systems, Inc., 2900 N. Loop West, Suite 1300 Houston, TX 77092 (800) 362-9118	PC/XT/AT, MS-DOS
Harvard Total Project Manager	Sofware Publishing Corp., P.O. Box 7210, 1901 Landings Dr., Mountainview, CA 94039 (415) 962-8910	PC/XT, MS-DOS
MACProject	Claris Corporation 440 Clyde Ave. Mountainview, CA 94043 (415) 960-1500	MAC, CP/M
Microsoft Project	Microsoft Corporation, 16011 NE 36th Way, Box 97017, Redmond, WA 98073 (206) 882-8080 (800) 541-1261	PC,PC/XT, MS-DOS
MICROTRAK	SOFTRAK Systems, 1574 W. 1700 S., Suite 2C, Salt Lake City, UT 84104 (801) 973-9610	PC,PC/XT/AT, MS-DOS AT&T 3B2, Unix PC, HP Series 2000, Tandy 6000, Sun Microsystems, UNIX/XENIX
PLOTRAK	SOFTRAK Systems, 1574 W. 1700 S., Suite 2C, Salt Lake City, UT 84104 (801) 973-9610	HP Series 7000 Plotter, Houston Instruments DMP Series Plotters, MS-DOS

Program	Vendor	Hardware/Operating System
		Calcomp 1042, 43, 44; Ioline LP3700, Roland RLDXY, 880, UNIX/XENIX
PERT MASTER	Westminister Software Inc., 3235 Kifer Rd., Santa Clara, CA 95951 (408) 736-6800 (800) 822-8298	PC,PC/XT, MS-DOS
PLANTRAC	Computerline, P.O. Box 308, 52 School St., Pembroke, MA 02359 (617) 294-1111	PC,PC/XT, MS-DOS
PMS-80	Pinnell Engineering, Inc., 5331 SW Macadam Ave., Suite 270, Portland, OR 97201 (503) 243-2246	PC,PC/XT, MS-DOS, AT&T 3B2, UNIX
Primavera Project Planner	Primavera Systems Inc. Two Bala Plaza, Suite 925, Bala Cynwyd, PA 19004 (215) 667-8600	PC/XT/AT, MS-DOS
Primavision	Primavera Systems Inc., Two Bala Plaza, Suite 925, Bala Cynwyd, PA 19004 (215) 667-8600	HP7475, 7580, 7585; Houston Instruments DMP Series Plotters; Nicolet Zeta 8, 822, 836; CalComp 104X, MS-DOS
PROB-PERT	SCS Telecom Inc., 107 Haven Ave., Port Washington, NY 11050 (516) 883-0760	PC/XT/AT, MS-DOS and LOTUS 1-2-3
PROMIS	Stategic Software Planning Corporation, One Athenaem St., Cambridge, MA 02142–9936	PC/XT/AT, MS-DOS

Program	Vendor	Hardware/Operating System
QUIKNET PROFES- SIONAL	(617) 577-8800 (800) 821-8346 Project Software and Development, Inc. (PSDI), 20 University Rd., Cambridge, MA 02138 (617) 661-1444 (800) 231-7734	PC/XT/AT, MS-DOS
SuperProject Expert	Computer Associates International, Inc., 1240 McKay Dr., San Jose, CA 95131 (408) 432-0614 (800) 533-2070	PC,PC/XT/AT, MS-DOS
Time-Line	Symantec Corp., 505B San Marin Dr., Novato, CA 94945-1310 (415) 898-1919	PC,PC/XT/AT, MS-DOS
VISION Micro	Systonetics, Inc., 1561 E. Orangethroupe Ave., Fullerton, CA 92631 (800) 854-1780	PC/XT/AT, MS-DOS
VUE	National Information Systems, 1190 Saratoga Ave., Suite 100, San Jose, CA 95129 (408) 985-7100	PC/XT/AT, MS-DOS
ViewPoint	Computer Aided Management, 1318 Redwood Way, Petaluma, CA 94952 (707) 795-4100	PC/XT/AT, MS-DOS
WINGS	AGS Management Systems, Inc., 880 First Ave., King of Prussia PA 19406 (215) 265-1550	PC/XT/AT, MS-DOS, MicroVAX, VMS, AT&T 3B, UNIX

The authors welcome further information regarding these and other available software programs.

APPENDIX 11.2 A REPRESENTATIVE LISTING OF MAINFRAME AND MINICOMPUTER PROJECT MANAGEMENT SOFTWARE VENDORS

Vendor	Software Systems
Automatic Data Processing 175 Jackson Plaza Ann Habor, MI 48106 (800) 521-3166	APECS/8000
AGS Management Systems, Inc. 800 First Avenue King of Prussia, PA 19406 (215) 265-1550	PAC II, PAC III, WINGS
Computer Associates 10505 Sorrento Valley Road San Diego, CA 92121 (619) 452-0170 (800) 468-0725 (within CA) (800) 841-3734 (rest of U.S.)	CA-TELLAPLAN
Metier Management Systems, Inc. 2900 N. Loop West Suite 1300 Houston, TX 77092 (800) 362-9118	ARTEMIS
National Information Systems 1190 Saratoga Avenue Suite 100 San Jose, CA 95129 (408) 985-7100	VUE
Primavera Systems, Inc. Two Bala Plaza Suite 925 Bala Cynwyd, PA 19004 (215) 667-8600	Primavera Project Planner, VAX Version

Vendor	Software Systems
Project Software & Development, Inc. 20 University Road Cambridge, MA 02138 (617) 661-1444	PROJECT/2
Shirley Software Systems 1936 Huntington Drive South Pasadena, CA 91030 (818) 441-5121	MISTER
Systonectics, Inc. 1561 E. Orangethourpe Ave. Fullerton, CA 92631 (800) 854-1780	VISION EZPERT
T and B Computing, Inc. 1100 Eisenhower Place Ann Arbor, MI 48104 (313) 973-1900	TRAKSTAR

Comments from individual readers concerning additional MIS systems are welcome. Please bring to the authors' attention the names of any software vendors whose product should be included in this list.

APPENDIX A

RISK ANALYSIS

The use of probabilistic PERT requires that engineering judgment be employed to determine the optimistic, pessimistic, and most-probable completion times of each activity. The difference between the pessimistic and optimistic times is proportional to the standard deviation, risk, or uncertainty in an activity's duration. The perceived risk or uncertainty in an activity or project should be justified, since what may be basic research in one company may be an "old hat" solution to another.

Figure A1 shows the criteria used for classifying *risk* as high, medium, or low. Note that cost, schedule, as well as technical criteria must be evaluated to determine the risk of an activity or project. Further, if any single category is, for example, high risk, the entire activity or project becomes high risk.

The use of Fig. A1 permits the engineering manager to focus on the specific aspects of an activity or project that causes it to be of higher risk than desired. In this way the risk can possibly be decreased.

Table A1. Criteria for Classifying Risks as High, Medium, or Low.

Risk	Cost	Schedule	Technical
H I G H	· Requires additional outside help of uncertain cost · No quotations received/exist · Lack of definition prevents estimating	· Extensive research studies required · Extensive tests and evaluation required · Extensive technical problems · Lack of proper personnel · Conflict with other programs · Tasks unique: never done before · Facilities to be located, constructed, or defined · Required personnel to be recruited · Vendors not used previously · Multiple vendors not available	· Beyond state-of-art · Research required · Never studied · No in-house experience · New technology required · Poor manufacturing technology · Design approach uncertain · Proper personnel availability questionable · No field experience

M E D I U M

- Within cost on similar programs
- Estimates are based on extrapolations

- Associated efforts started
- Prototype development in process
- Previous achieved schedule improvements planned
- Adaptable hardware, software, etc
- Some facilities to be modified
- Proper skills identified—recruitment assignments planned
- Vendors are available, but have not been used

- At state-of-art
- Research in progress
- Capabilities available
- Adaptation of existing capability
- Manufacturing technology being implemented or planned
- Some prior experience

L O W

- Past history on similar jobs
- Under cost on similar programs
- Firm quotations received
- Multiple sources exist

- Past history on similar jobs
- Existing hardware/software computer programs, management systems, etc.
- Previously performed similar tasks within or below projected schedule
- Facilities/capacity committed
- Proper personnel assigned
- Previous vendor histories satisfactory

- Applicable research completed
- Technology demonstrated
- Proper personnel committed
- Extensive experience
- Prototype developed
- Already Met RFP requirements
- User as well as manufacturer
- Relevant performance data (empirical) exists

INDEX